Agriculture

Improving Agricultural Education
Concept and Methods

Improving Agricultural Education
Concept and Methods

Editors
Birendra Kumar, *Ph.D.*
Shivendra K. Kashyap, *Ph.D.*

BIOTECH
2012
BIOTECH BOOKS®

ISBN 978-81-7622-256-3

The views expressed in various chapters are those of the author and not of editor or publisher of the book.

Published by: **BIOTECH BOOKS®**
4762-63/23, Ansari Road, Darya Ganj,
New Delhi - 110 002
Phone: +91-011-23262132
E-mail: biotechbooks@yahoo.co.in

Printed at: **Chawla Offset Printers**
Delhi - 110 052

PRINTED IN INDIA

Preface

Agricultural education has assumed special significance in the twenty first century in the wake of myriad socio-economic and technological changes nationally and internationally. The challenges of global economy have affected the way education be imparted, choice of curriculum, innovative methods of delivery and approaches to train breeds of professionals facing new problems and dealing with situation needing interdisciplinary connections. There is need for educationists sensitive to the emerging problems and knowledge about educational methods as well as technical subject matters. Such specialists are scarce who reflect over their teaching and professional practice critically. The book an edited volume, fills the gap by inviting contributions from experienced academics and synthesizing their inputs into a coherent whole. We are sure this will provide stimulus to many more academics to come out. The untiring efforts of education division of the Indian Council of Agricultural Research as well as leading SAUs to bring reforms in agricultural education have brought tremendous changes in learning. There is need for collecting experiences and experiments to give scientific vigor to the practice of agricultural education. We shall welcome the feedback from our valued readers.

We remain grateful to our respective organization(s) for all the support and stimulus to pursue our writing project.

Birendra Kumar, *Ph.D.*

Dean, PGS
Bihar Agril. University
Sabour, Bhagalpur
Bihar
E-mail: bi.kumar@yahoo.com

Shivendra K. Kashyap, *Ph.D.*

Associate Professor
Agricultural Communication
GBPUAT, Pantnagar
Uttarakhand
E-mail: kashyapsk@rediffmail.com

Contents

List of Contributors v

Preface v

Section 1
Education and Agricultural Education

1. Indian Philosophy of Education: Emerging Challenges and Contemporary Issues 3
B. Mohan Kumar

2. Agricultural Higher Education in India: A Historical Perspective 14
Milind Ahire, Birendra Kumar and Aparna Rajput

3. New Trends in Agricultural Education 23
V.P.S. Arora

4. Higher Agricultural Education in India: Issues and Challenges 30
B. Kumar

5. ICT Applications in Agricultural Universities 51
Subir Ray

6. Value Added Agricultural Education: Issues and Strategies — 62
Birendra Kumar and Renu Jethi

7. Excellence in Agricultural Education: A Case of Pantnagar — 66
Birendra Kumar and Aparna Rajput

Section 2
Educational Technology

8. Educational Technology: Concept — 77
S.K. Kashyap and Birendra Kumar

9. Educational Technology in State Agricultural Universities: Past, Present and Future — 80
Birendra Kumar and S.K. Kashyap

10. Education Technology Cell: Concept and Future Outlook — 85
S.K. Kashyap and Birendra Kumar

11. Experiential Learning in Teaching — 88
Neelam Bhardwaj and S.K. Kashyap

12. Agricultural Education through Distance Learning — 99
Birendra Kumar and Aparna Rajput

Section 3
Improving University Teaching

13. Professionalizing Teaching in Agricultural Universities: Need and Strategy — 115
Birendra Kumar

14. Learning Theories and their Implications in Teaching — 126
Neelam Bhardwaj

15. Skills in Teaching — 136
Birendra Kumar and S.K. Kashyap

16. Methods of Teaching 143
Birendra Kumar and S.K. Kashyap

17. Use of Teaching Materials and Aids 162
S.K. Kashyap and Birendra Kumar

18. Games and Exercises for Enhanced Learning 181
Shreena Keshwani and Richa Lohani

19. Role Play in Method of Teaching 194
Surya Rathore

20. Teaching Packages for Improving Quality of Teaching 202
S.K. Kashyap and Birendra Kumar

Section 4
Good Teaching Practices: Issues and Experiences

21. Application of Total Quality Management in Home Science Teaching 209
Manisha Gahlot and Neema Papnai

22. Veterinary Education in India: Issues and Strategies 216
R.J. Sharma

23. Quality Teaching of Agronomy: A Reflection 223
M.S.Pal

24. Successful Teaching of Plant Pathology 230
S.C. Saxena

25. Experiences in Teaching Plant Pathology 237
S.J.Kolte

Section 5
Communication Skills for Effective Teaching

26. Creating Motivation in Class Room: Some Tips 247
Birendra Kumar and S.K. Kashyap

27. **Personality Development of Teachers** **366**
S.K. Kashyap and Birendra Kumar

Section 6
Evaluation as a Tool of Teaching

28. **Evaluation in Teaching Learning Process** **277**
Renu Jethi and Birendra Kumar

29. **Designing Tests and Evaluation Approaches** **284**
A. K. Singh and Surya Rathore

30. **Training of Teachers for Qualitative Improvement in Teaching** **292**
Birendra Kumar and Renu Jethi

Bibliography **299**

List of Contributors

Dr. A.K. Singh

Former Associate Professor,
Department of Agricultural Communication,
G.B. Pant University of Agriculture & Technology,
Pantnagar – 263 145, Uttarakhand

Presently, Professor
Agricultural Institute,
Banaras Hindu University

Dr. Renu Jethi

Scientist
VPKAS
Almora, Uttarakhand

Dr. Aparna Rajput

Research Scholar
Department of Agricultural Communication
G.B. Pant University of Agriculture & Technology,
Pantnagar – 263 145, Uttarakhand

Er. Richa Lohani

Business Analyst
Genpact, Gurgaon

Dr. Birendra Kumar

Former Director Communication
G.B. Pant University of Agriculture & Technology,
Pantnagar – 263 145, Uttarakhand

Presently, Dean PG Studies
Bihar Agricultural University
Bihar

Dr. R.J. Sharma

Former Dean,
College of Veterinary & Animal Sciences
G.B. Pant University of Agriculture & Technology,
Pantnagar – 263 145, Uttarakhand

Dr. B. Mohan Kumar

Professor & Head,
Department of Social Sciences & Humanities,
College of Basic Sciences and Humanities,
G.B. Pant University of Agriculture & Technology,
Pantnagar – 263 145, Uttarakhand

Dr. S.C. Saxena

Ex- Professor
Department of Plant Pathology
G.B. Pant University of Agriculture & Technology,
Pantnagar – 263 145, Uttarakhand

Dr. Mahendra Singh Pal

Associate Professor
Department of Agronomy
G.B. Pant University of Agriculture & Technology,
Pantnagar – 263 145, Uttarakhand

Dr. S.J. Kolte

Ex- Professor
Department of Plant Pathology
G.B. Pant University of Agriculture & Technology,
Pantnagar – 263 145, Uttarakhand

Dr. Milind Ahire

Assistant Professor
Mahatama Phule Krishi Vidyapeeth
Maharashtra

Dr. S.K. Kashyap

Associate Professor,
Agricultural Communication
and Asssociate Director Communication,
G.B. Pant University of Agriculture & Technology,
Pantnagar – 263 145, Uttarakhand

Dr. Manisha Gahlot

Professor
Department of Clothing & Textiles,
College of Home Science,
G.B. Pant University of Agriculture & Technology,
Pantnagar – 263 145, Uttarakhand

Prof. Subir Ray

Ex-Professor and Head
Electrical Engineering Department,
G.B. Pant University of Agriculture & Technology,
Pantnagar – 263 145, Uttarakhand

Neema Papnai

Assistant Professor
Mody Institute of Technology and Science
Rajsthan

Dr. Surya Rathore

Associate Professor
Department of Agricultural Communication
G.B. Pant University of Agriculture & Technology,
Pantnagar – 263 145, Uttarakhand

Dr. Neelam Bhardwaj

Professor & Head
Department of Agricultural Communication
G.B. Pant University of Agriculture & Technology,
Pantnagar – 263 145, Uttarakhand

Dr. V.P.S. Arora

Former Dean College of Agribusiness Management,
G.B. Pant University of Agriculture & Technology,
Pantnagar – 263 145, Uttarakhand

Presently, Vice-Chancellor,
Kumaon University,
Uttarakhand

Section 1
Education and Agricultural Education

Chapter 1

Indian Philosophy of Education: Emerging Challenges and Contemporary Issues

There are myriad aspects of education that have moorings in a sustained tradition of philosophical thoughts. The institutional forms and substantive contents at various levels of education have been shaped by a blend of philosophical ideas and pragmatic needs of the times. There has been a splendid journey from the sacred and ritualistic to secular and rational, from conservative to progressive, and faith-defined contents to scientifically proven ones. The pious commitment of teaching has undergone a change to acquire the position of a professional venture. The prime focus of majority of students may have become securing a parchment of paper certifying their credentials rather than committed and arduous learning and honing potentials as wholesome beings. The mushrooming of substandard institutions has further compounded the problem. But there has remained philosophical grounding of educational system in old civilizations and societies like India that continues to provide a positive direction and optimistic vision.

Basic Tenets of Indian Philosophy

The word philosophy is of ancient Greek origin: *philosophía*, meaning 'love of wisdom'.

The term Indian philosophy subsumes several traditions of philosophical thoughts that originated in the Indian subcontinent. These include Vedic and its numerous philosophical off-shoots emerging through accretion and reformulation based on the *Vedas, Upanishads, Dharmshastras, Smritis, Epics, Puranas* and the six main philosophical systems. There also emerged Buddhism, Jainism or other philosophies through intra-systematic differentiation; fission in tradition and diversification of forms each having numerous categories and sub-sets. The inherent normative principles have been as systematic and rather closed as its empirical structure has been varied and amorphous or even eclectic. Behind the extremes of permissiveness and variety in ritual practices, there exists a logical view of order and change, being and becoming, creation and destruction and spiritual transcendence.

Having the same or rather intertwined origins, all of these philosophies have a common underlying theme related to *Dharma* (normative order), *Guna* (attributes) and *Karma* (personal moral commitment) among others. An attempt to explain the attainment of self-realization and *Nirvana/Moksha* as ultimate goal can also be discerned. These have been formalized and promulgated since several centuries with varied interpretations, commentaries and reformations continuing up to now. In the history of the Indian subcontinent, following long processes of amalgamation and syncretism between indigenous and exogenous cultures, the development of philosophical thought over a period of millennia gave rise to what came to be formalized the six systems or schools of *Aastika* or theistic (believers in monism or polism) and three of *Nastika* or atheistic philosophy. In the former category, there are *Sankhya, Yoga, Nyaya, Vaisheshik, Mimansa* and *Vedanta* philosophies, while Buddhist, Jain and *Charvaka* are in the latter one. These have shaped a diversity of educational patterns and systems in Indian society.

Traditional Education

Often the educational responsibility used to remain vested with the religious or philanthropic organizations besides established *Peethas, ashramas* or schools. The traditional *Gurukuls* were the main educational institutions located in places away from settlements. Some of the universities like the renowned universities of Takshashila and Nalanda were exemplary institutions blending the Buddhist and ancient accumulated knowledge. The focus had remained on wholesome personality development including sound health, ethically defined character, gradual awareness and proficiency in formal knowledge, functional skills and improved aesthetic sense among others. These were interwoven in a disciplined and demanding *Guru-Shishya* linkage having in-built expectations of dedication, perseverance, subservience and humility on the part of the learners; and intellectual brilliance, teaching excellence, lofty morality and objective discerning capacity embedded in the personality of teachers.

The last century witnessed a resurgence of revival in the traditional educational system blended with cotemporary advances in various forms. For instance, the *Gurukuls* were revived by Dayanand as a part of the Aryasamaj reform movement to reassert the supremacy of Vedic knowledge and sensitize to the superior notions. Ramkrishna Mission initiated by Vivekanand has been excelling in such education with a difference. Their popularity in some regions is credited to simplification of complex ideas and practices. Such movements bolstered the education system. However, the discerning trends among most of the existing traditional institutions have been to transform into seats of modern education to a greater extent.

Similarly indigenized versions of Christian and Islam philosophies with their sect variants have greatly influenced the variegated Indian philosophy and education. The Islamic educational institutions such as religious seminaries or *Madarsa* have been imparting traditional education though their emphasis continued to be more on religious, ritual and sacred issues and less on modern secular ones. However certain Islamic institutions having thrust on high professional education have substantially evolved into seats of excellence rather than remaining confined to scriptural and ritual instructions. The Christian organizations have been providing yeomen service in the area of modern education especially through public schools and colleges established all over even in remote hinterland to benefit underprivileged as well though the earlier initial objectives geared towards proselytization and formation of service functionaries.

Thus, great diversity in educational thought and practice has been nurtured by the liberal universalism embedded in the Indian philosophy. These diverse philosophical traditions have shaped the inherent ethos, styles, strategies, contents and objectives of education in varying temporal and spatial formations.

Moral Dilemma

With the prevalence of democratic ideology, desacredized secular ethos and materialistic-individualistic modernity; the education system based on scientific and professional requirements has come to be accepted as the most appropriate and desirable form. The thrust of modern education on acquiring professional competence, employable skill and scientifically defined rationalism has paved the path of unprecedented development in a span of a millennium or so; but the relative indifference to aesthetic and moral issues, the social and national commitments may have created a moral vacuum among the educated community. Many of the modern day problems of deviance, violence and degeneration can be traced to the falsified notion of considering education a panacea even if the inculcation of values is getting marginalized and material-monetary considerations gaining predominance.

Purposes of Education

The philosophy of education is concerned to the exposition of the purposes, nature and ideal content of education. Related topics include knowledge itself, the knowledge-provider and knowledge-seeker learner, values defining education, problems of educational management, and the relationship between education and society.

Two goals of education are enshrined in the texts of Indian philosophy: esoteric (*Para*) and utilitarian (*Apara*), the former related to spiritualism and the latter to materialism. These are also labeled as *Vidya* and *Avidya* that should be learned together for proper conduct of moral and social obligations. One can manage the worldly affairs by excelling in *Avidya* and acquire immortality (*Amritatva*) through the *Vidya*. Originally, education meant the knowledge of the Vedas but subsequently it also came to include the learning of various artistic, scientific and linguistic capabilities. The preset day prime concern has got focused only on the worldly or materialistic knowledge at the cost of spirituality, thereby disturbing the balanced existence of humanity.

It may be pertinent to exposit the notion of goal-orientations, the four *Purusharthas* as encapsulated in the Indian philosophy:

Dharma (goals of moral obligation), Artha (the economic-utilitarian goals), Kama (aesthetics, sex, and other material goals of sensory enjoyments) and Moksha (the pursuit of salvation from the chain of birth and rebirth). These have relevance in the contemporary domain of existence having plenty of life-choices and opportunities, though not evenly dispersed and made easily accessible. We can posit somewhat synchronous purposes of education in different terms succinctly proposed as follows:

Sensitization

It is one of the basic purposes meant to expose and motivate people towards the path of righteousness from the very beginning. In fact, the society depends on educating people to become healthy, responsible, sensitive and enterprising. This is an intricate, challenging task requiring deep understanding of ethical principles, moral values, aesthetics, socio-political and economic issues, besides rational domain of techno-scientific issues.

Enablement

Education can develop skills and capacities on which progress in every practical field depends. It is thus a means to foster development and prosperity of the individuals, society, and even humanity.

Accomplishment

Education tends to provide a firm foundation for the achievement of personal fulfillment. Individual development and the capacity to fulfill purposes can depend on an adequate preparation during the course of acquiring education.

Self-realization

This pertains to loftier goal of life ideally enshrined in the religious and philosophical texts. It relates to endowing persons with adequate discerning capacity to opt for virtuous acts through restraint and sacrifice, and gradually attain detachment leading to *Nirvana*.

Continuum of Knowledge

A central tenet of education typically includes "the imparting of knowledge." At a very basic level, this purpose ultimately deals with the nature, origin and scope of knowledge. The branch of philosophy that addresses these and related issues is known as epistemology. This area of study often focuses on analyzing the nature and variety of knowledge and how it relates to similar notions such as truth and belief.

While the term, knowledge, is often used to convey this general purpose of education, it can also be viewed as part of a continuum of knowing that ranges from very specific data to the highest levels. Seen in this light, the continuum may be thought to consist of a general hierarchy of overlapping levels of knowing. Students must be able to connect new information to a piece of old information to be better able to learn, understand, and retain information. This continuum may include notions such as data, information, knowledge, wisdom, and realization.

Holistic Education

One of the ideals of Indian philosophy is to focus on holistic education. It is conscious transformation that aims at holistic and simultaneous welfare of one and all. This requires conscious development of fitness of one's body, refinements of instincts, broadening and profoundness of emotions, blossoming of intelligence and liberating perspective of universal oneness. The objectives of education do not remain confined to acquiring formal certificates or degrees, but taken as a complex process of experiential learning, improving capabilities and getting enabled to secure compatible work. It amounts to exploring the life-chances and opting for the best of life-choices.

Some of the inherent positivistic and pragmatic implications of such education may be as follows:

- ☆ To maintain psychosomatic fitness;
- ☆ To be inquisitive and develop a sense of inquiry;
- ☆ To realize one's potential with sustained efforts;
- ☆ To keep on honing, upscaling and excelling skills;
- ☆ To accept that success can be secured through
    - Setting an achievable goal,
    - Planning based on resources and skills,
    - Self-discipline and perseverance
    - Grooming and enhancing adequate linkages/ networks

- ☆ To realize that one-time or successive failure(s) is a sort of deferred success that demands concerted and patient efforts.
- ☆ To condition oneself to treat others the way one would prefer to be treated by others. In fact, this happens to be the basic surmise of Indian Philosophy.

Emerging Issues in Education

The age-old and ever-evolving educational issues are continually being studied, investigated, critiqued, discussed and modified by researchers, politicians, and educators themselves. Even a cursory reading of educational history reveals about multiple progresses. One of the remarkable achievements is that education has moved from the privilege of an insignificant portion of the population to the prerogative of all. Students have been able to pursue their education for longer periods of time. The scope of curriculum has expanded in ever-emerging sunshine areas and directions, and progress has been made in the educative process itself. The art of teaching, once thought to be a mysterious "gift," is now something to be studied and taught.

Despite this advancement, there remains great progress to be made and significant challenges to be met. A recurring theme throughout the history of education is equal access to education by all. In spite of the improvement seen in the education of large numbers of people worldwide, there remain areas where illiteracy predominates and drop-out rate persists.

Presumptions about Education

When most of us think about higher education, we tend to make numerous assumptions. These are related to various principles, issues and requirements in education. Given the diversity and wider scope, it is not possible to bring into focus all presumptions. For learning, the course is often the basic unit of measure that we presume. Programme length is measured in semesters or years. Courses originate from within, rather than outside, the institution. These need to be reassessed and refocused to gain maximum advantage.

Some other common systemic assumptions about higher education may include:

- ☆ The students' profile and learner preferences are taken to be known for learning and service delivery.
- ☆ Student credit hours and full-time equivalents are relevant units of measure in higher education.
- ☆ Programmes and Courses being offered are adequate to satisfy needs of students.
- ☆ Completion of the curriculum is the measure of competency.
- ☆ Traditional institutional models (*e.g.*, for classroom instruction, governance, and financing) will be successful in an e-learning world.
- ☆ Higher education must provide all components of the educational process (*e.g.*, content, curriculum, services, and credentialing).
- ☆ External providers of educational services (*e.g.*, courses or tutoring from independent individual or an Internet start-up) are inherently bad or of lower quality than educational institutions.
- ☆ Quality is better in a not-for-profit institution than in a for-profit one.
- ☆ High quality will drive out low quality.
- ☆ The faculty member is the focal point of the learning process in which feed-back from the learners is not essential.

Most of these assumptions are facile, untenable and unrealistic given the currently emerging challenges in education and expectations of stakeholders.

Constraints and Challenges of Education

For India to achieve its target of growth and powerful status, it needs to harness all its resources–the biggest one being manpower. But it is barely equipped to take on the challenge of leading India to become a power to reckon with. About 36% of the population is illiterate and the ones, who are certified literate, are barely employable. This is because their curiosity to learn and discover their unique potential is dulled and stamped out by rote learning and testing approach. Everybody has to learn the same way, is tested the same way and there is no accommodation for each child's uniqueness. It is baffling as to how the inherent differentiation in background and socialization resulting into varied intellectual capacity and inclination can be straight-jacketed in uniform patterns of education with the expectation to access and excel in formative periods against all odds!

The diversity in educational institutions, their concerns and resources available does not allude to easier identification of constraints and challenges. Only certain general issues can be raised as instances:

- People in more remote or agrarian areas are sometimes unaware of the importance of education. Even if they become aware, their resources and priorities impede their access to secure education.
- There is also economic pressure from those parents who prefer their children making money in the short term over any long-term benefits of education. But when poor families reach a certain economic threshold where families are able to provide for their basic needs, parents return their children to school.
- There is a lack of world class universities and other educational institutions particularly in resource-scarce regions.
- There are uniform, over structured, inflexible centralized programmes often from a central agency that regulates all aspects of education without taking into account the inherent differences and requirements.
- According priority to memorization over comprehension and patterned thinking over innovative or creative one tend to diminish the spirit of inquiry and commitment.
- Despite inclusion of sunshine areas and state-of-art technology in certain disciplines, the outdated styles of instruction persist disenchanting students and other stakeholders.
- Due to globalization, there is increased pressure on students in curricular activities and acquiring competitive edge primarily in programmes having increasing demand in the market.

- The recession, if it persists, may make certain technical-professional education somewhat less important and general employability scarcer.

There are certain other issues that need to be highlighted. For example, the education of children who are not "typical" students occupies many of the efforts of educators. These students include those who cannot or do not make educational progress at the rate of other children; make progress much faster than their same-age peers; speak a language other than the language of instruction; and are younger or older (*e.g.*, adult education students) than traditional age students.

Digital Educational Technology

While research issues may have been facing educators for decades, the nature of scholarly communication is being transformed both in form and content. The nature of the medium alone allows researchers, academics and practitioners to disseminate ideas in a timely manner through a dynamic medium that offers an array of possibilities in terms of presentation of ideas and information (*i.e.*, enhanced opportunities for visual, audio and video interactivity, on-line discussion possibilities, convenient links to other resources). We, as a community of scholars, are thus challenged to think about our ideas and their presentation in new and innovative ways.

India is now developing technologies that will skip land based phone and internet lines. India launched EDUSAT, an education satellite that can reach more of the country at a greatly reduced cost. It would be imperative to provide computer equipment, learning materials and internet access. More compatible and curricula-oriented software are required to facilitate adequate use of computer-added technology. The initiative by a group out of MIT and major corporations to develop a cheaper laptop costing $100 or about Rs5000 may enable to give children a digital education, and to close the digital divide.

The burning issue is how we can ensure sustained functioning of such technology in the face of energy crisis and resource crunch besides training and motivational deficiency among the educators and learners!

Implications

As the information age progresses at an unprecedented rate, national and global boundaries continue to dissolve. Thus, we seek work that addresses both national and international issues relevant to the field at large in the knowledge that we are a global community of scholars with much to learn and much to share.

- ☆ There is a big gap today between what education needs to do and what it is actually doing.
- ☆ To convert manpower into valuable human capital, education needs to help students achieve their true, unique potential.
- ☆ They need to be equipped with life skills such as critical thinking, conceptual understanding and strong moral fiber to succeed in the new knowledge economy.
- ☆ They have to be prepared to face any test in life through practice and wide exposure.
- ☆ The education policymakers, educators and administrators need to come out of slotted conventional policies, styles and strategies; and replace these with innovatively designed ones incorporating needs, demands and preferences of the times.

Chapter 2

Agricultural Higher Education in India: A Historical Perspective

Higher education in India is nearly four centuries old; the Jesuit college in Goa was perhaps the first to start in 1575 and another college in Baroda the same year. The first three Indian universities at Madras, Bombay and Calcutta were established in 1857. Several decades later, early in twentieth century, when six agricultural colleges were established in undivided India at Kanpur, Lyallapur, Coimbtore, Nagpur in 1906 and at Poona and Sabour in 1907 and 1908 respectively. In 1914, the Allahabad Agricultural Institute was established through the efforts of Dr. Sam Higginbottom. On the recommendations of Royal commission of Agriculture (1928) the Imperial Council of Agricultural Research (ICAR) was established in 1929, which was renamed after independence as Indian Council of Agricultural Research on June 10, 1948. The ICAR was originally started as a central agency mainly for co-coordinating activities associated with Agriculture. At the time of independence in 1947 there were 15 institutions in the country providing agriculture and veterinary education with total in take capacity of 1500 students. Soon after independence, University Education Commission headed by Dr. S. Radhakrishnan brought out in its report 1949 the need for orienting Agriculture Education in India towards solving farmers' problems and recommended major modifications in curriculum for higher education in Agriculture so as tolerate such an education to rural development. The term higher agricultural education refers to under-graduate education leading to Bachelor's degree and two levels of post-graduate education leading to Master's and Doctoral degree in specialized subjects. Agriculture is generally used to connote crop production, Horticulture, Veterinary Science and Animal Production, Agricultural Engineering, Home Science, Fisheries, Forestry, Food Technology, etc. Before 1960, Agriculture and veterinary colleges operated in the different states of India as affiliated colleges of general universities which had jurisdiction for higher education in liberal arts, humanities, social science, natural sciences and several professional areas like medicine, engineering, agriculture etc. Veterinary and Animal Husbandry colleges were run by the respective State Governments. However, the agricultural colleges were operated both by State Governments and a few private organizations. A few home science colleges were administered by the State Governments directly. Only one college under private management (Allahabad Agricultural Institute) awarded a degree in Agricultural Engineering initially, while, much later, the Indian Institute of Technology, Kharagpur, recognized as an Institute of national importance, introduced this degree programme along with other branches of engineering. After independence the Central Government was concerned with the need to bring about changes in higher education in agriculture to make it more relevant to the emerging needs of agricultural development. In the search for a model, the United States Land Grant College System attracted attention. Then onwards the efforts to establish agriculture university were taken on the recommendations of various committees and experts as detailed below:

- ☆ The first – joint Indo–American team headed by Shri K.R.Damle (1955) endorsed the recommendations of 'University Education Commission' and recommended one Agriculture University in each state.
- ☆ In 1955 Dean Harold Hannah came to India and spent two years in developing a "Blue print for an Agricultural University". His report and recommendations were accepted as the basis for development of University and the U.P. Government set up a "Development committee" to prepare plans and estimates.

- ☆ Dean Harold Hannah returned to India on a three month assignment in 1959 for further consultation and finalization of plans for establishing the university. The contract under which the University of Illinois began rendering technical advice and assistance to the U.P. agricultural University was approved by the Board of trustees of the university on Oct 30, 1959.
- ☆ Uttar Pradesh Agricultural University located in Nainital district on an area of about 16,000 acres.The first registration of new students in the university, however, was held on July9, 1960, and class work began on July 11 with an enrolment of 246 students(145 in college of Agriculture and 101 in College of Veterinary sciences). The university was later dedicated on november17, 1960, to late Pandit Jawaharlal Nehru, then Prime minister of India.
- ☆ In 1960, the second joint Indo- American team headed by Dr M.S. Randhawa manifested further interest in establishing more Agriculture Universities in different states came up. The ICAR set up a committee in 1961 called the "Agriculture University Committee: in order to render technical assistance to all the interested states in establishing agriculture University.

  In 1962 expert committee appointed by Government under leadership of Dr R. W. Cummings reported their as below
- ☆ In an Agricultural university, the teaching staff, in addition to having responsibility for resident instruction, will also have a responsibility for applied as well as fundamental research in agriculture.
- ☆ The curricula and the training programmes in Agricultural University are modeled in keeping with the needs of the state on one hand, and with the aptitude and needs of the students on the other; this means that curriculum is flexible and is capable of being offered in form of specialized courses covering a wide ground of subject matter.
- ☆ After 1963 more emphasis was given on expansion and consolidation of Agricultural education.
- ☆ The Education Commission under the chairmanship of Dr. D.S.Kothari (1964 – 1966) was another landmark in

the history of agricultural education in the country. The Commission stressed the importance of agricultural university in each state. All the states have adopted this concept and now almost all the states have at least one agricultural university to fulfill the regional needs.

☆ The model act given by Dean Hannah was modified in 1981, which prescribed the following main objectives of an agricultural university:

1. Making provision for imparting education in different branches of study, particularly agriculture, horticulture, veterinary and animal sciences, fisheries, forestry, agricultural engineering, home science and other allied branches of learning and scholarship.
2. Furthering and advancement of learning and prosecution of research, particularly in agriculture and other allied sciences.
3. Undertaking the extension of such services to rural people of the State.
4. Such other purposes as the university may determine from time to time.

The Indian Council of Agricultural Education established under the overall control of ICAR was charged with the responsibility of brining about greater standards of uniformity in the agricultural curricula and serve as forum for discussion of problems and issues concerning higher agricultural education within a national perspective. In this period of Second Educational Commission recommended that the function of co-ordination and promotion of higher Agricultural education could best be carried out by ICAR. It is also suggested that the setting up of UGC type body or higher agricultural education, and the standing committee for Agricultural Education set up under the reorganized ICAR largely performed the role until 1973.When ICAR underwent the second major reorganization in 1973, this standing committee was abolished and most of its technical function were entrusted to the scientific panel on Higher Agricultural Education. At present the Indian Council of Agricultural Research (ICAR) is the national apex body on agricultural education and research. It has responsibility to plan, undertake, guide, support, promote and coordinate agricultural education (besides research). ICAR thus plays the same role as the University Grants Commission (UGC) does for general universities in promoting higher education in agriculture and allied fields. However, unlike UGC, it does not enjoy any statutory authority to determine and enforce requisite standards of agricultural education imparted by the State Agricultural Universities (SAUs) and other Institutions (constituent faculties and colleges of Central Universities and private colleges affiliated to General Universities).

The Indian Council of Agricultural Research has been alive the needs of teachers' training since establishment of first few SAUs in the country. A systematic effort to enhance instructional skills of teacher was evident with the establishment of center of advanced study in educational technology at National Academy of Agricultural Research Management (NAARM) Hyderabad in 1979. Efforts have been initiated to establish regional centers for training of SAUs teacher's in Education Technology of late in 1994, a full-fledged institute has come up at Chaudhary Charan Singh Haryana Agricultural University; Hissar.

- ✰ The ICAR had launched an innovative AHRD project with the financial support of the World Bank during 1995. The focus of the project is to improve the quality and relevance of higher education and to strengthen the capacity of participating states to develop and manage their agricultural human resources.In the first phase it covered Tamil Nadu, Andhra Pradesh, and Haryana and one veterinary university of Tamilnadu with the Education Division of Indian Council of Agricultural Research as the coordinating agency at national level. Some of the significant achievements during phase-I of the AHRD project include; constitution of accreditation board to establish norms and standards for higher agricultural education; uniform academic regulations, examination and evaluation system in SAUs, revision of U.G. course curriculum; introduction of Rural Agricultural Work Experience (RAWE), new courses in computer application, bio-statistics, bio-diversity, bio-technology, agribusiness management, improvement of faculty competence, provision of sabbatical rules.
- ✰ Accreditation board was established in 1996 by ICAR. Accreditation board assesses institution with reference to

predetermined criteria. Accreditation board functions as in house unit of the ICAR restricting its activities only to those institutions that are established and or funded by ICAR. The implementation of recommendations are monitored by Education Division and reported to board annually. The accreditation status is valid for a period of five to ten years. The first accreditation visit was organized by Accreditation board during November 1998 and result was announced in August 1999.

The Challenges of Change

India has developed strong agricultural education system since independence. The establishment of SAUs, on the pattern similar to that of land-grant colleges of the united states, made landmark in reorganizing and strengthening agricultural education system The green revolution, with its impressive social and economic input would have not been possible without the contribution of these centers of learning's in the form of development of trained scientific manpower, the generation and assessment of new technologies and their dissemination to the farming community.

During the early part of establishment of SAUs the academic programmes were structured to produce graduates who became primarily technology agents. In the present era of specialization and development in modern science it is necessary that we restructure our agricultural education in a manner that the graduates coming out are analytical technology oriented, sensitive to social and economic issues of farming in India, and able to meet challenges of new millennium with determination, zeal and commitment. For this happen universities have to develop and strengthen both formal and non formal degree programmes brings academic excellence and education relevant to future needs.

College teachers of agriculture are often selected without considering their ability to teach.After being selected it is not common practice to give them any training in teaching. A frequent practice is to select young men who have made high grades in college and to make them assistant instructors or full instructors (Sort of assistant professor in rank) They may begin teaching without any instruction in the learning process, philosophy of education or method in teaching. They are judged by research and publication they do, while the quality of their teaching is ignored. If they show enough attitude in research they retained as instructors. The practice of rewarding teaching tends to create an indifferent attitude toward teaching on the part of college personnel. It handicapped the improvement of instruction.

In the last three decades, all the committees appointed to suggest reforms in Indian higher education have recommended the regular evaluation of teacher's performance and ensuring their accountability. In 1971, the S.R.Sen committee, while recommending the higher pay scales has added the need for code of conduct (professional ethics) to be made part of it. Pay scales were revised but this part was not been implemented. In 1986 National Policy of Education had recommended for 'Annual performance appraisal' of the teachers of college and universities. It suggested a 360 degree feedback, which included evaluation from self, peers heads of institution, department and others. In 1987, Mehrotra committee too stressed the need of compulsory annual submission of 'performance appraisal' (an assessment of performance of teacher's which would encourage their accountability) In December 1998, UGC issued a notification regarding "Accountability in Higher Education" for all the universities. Emphasizing the need for accountability in teaching profession. The Rastogi committee has suggested 'self appraisal by teacher's, assessment by students in appropriate manner, periodic performance appraisal having regard to number of teaching days, workload and code of professional ethics.

People in past thought that the teacher are born not made. It meant that any one could teach well if he knew his subject. The expansion of education consequent on industrialization and social advancement made it necessary to adopt methods of teaching large classes and gradually a body of techniques evolved. Training in these techniques, it was hoped would turn the average teacher in to good teacher and good teacher into better teacher.

In some cases such as GBPUA&T, Pantnagar, the universities at Ludhiana, Hissar and others, faculty expertise is comparable to what may be considered the best for standard. Training of faculty was greatly aided during the decades of sixties through various faculty developments and exchange programmes under the auspices of US Technical Co-operation Mission and later US agency for international development. Six American land grant universities participated in exchange programme with eight universities.

Appraisal of teaching has emerged in the changing perspective of education. The new education system has emerged operating on the principle of 'Consumerism' with the emphasis on 'Consumer satisfaction' 'Cost effectiveness' and 'teacher accountability' A good teacher need have no fears about assessment by any one, because teaching is joy for him.

Quality Improvement of Agricultural Education

Of late there has been widespread consensus that quality of graduates in various disciplines of agricultural sciences does not match with the changing needs. Hence there is urgent need for SAUs to give attention for upgrading the quality and standards of higher agricultural education. Academic in breeding is considered as major problem in most of the SAUs. Teaching method and curricula are to be modified, physical facilities, equipment and teaching aids have to be modernized and competency of teacher needs to be improved.

Qualities and Competence of Teachers

Even with the renewed interest in teaching during the past decade, most new professors remain, learn to teach by trial-and-error.

Student-teacher interaction needs greater attention through effective advisory system. Training of teachers has never received adequate attention in past. Assessment of quality Teaching and teachers should be a regular feature for improvement for attaining standards and excellence in Agricultural Education. Teacher talents and capabilities are directly correlated with the performance of students in term of their knowledge and skills development. A well defined and transparent career management and advancement policy need to be evolved for faculty in order to motivate them to enhance the job satisfaction and eliminating stagnation.but professional accountability and academic standards are to be maintained for attaining academic excellence of an Agriculture University.This could be possible by team work and mindset for the teachers and scientists by exploiting their hidden talents and capabilities towards common goal of academic excellence.

Initial Training of Teachers

Agricultural Teachers in Agricultural colleges and Agricultural University are selected and employed mainly on the basis of their agricultural professional degree and experience. This is not enough for effective teaching and therefore teachers should be trained through continuous in service training by way of strengthening their teaching methodology as well as training them in their subject matter to cope up with the current science and technology developments. Further more, the beginner agriculture teachers have not received any formal pedagogical preparation and therefore they need initial training at beginning of their job.

Continuous Training of Teacher

Learning a lifelong and continuous process is very important for everyone especially for teachers. The type and forms of learning may be varied like self directed learning,participate in distance learning attending workshop, seminars, symposia, etc based on requirements of teachers. Continuous training of the agriculture teachers and continuous learning are quite essential to update their knowledge on the emerging innovation in Agricultural sciences and also equip with the modern technologies.

Quality of teaching is not uniform all over the country. In order to flourish agricultural education, there is need to learn from Agricultural University Teachers while teaching what methodologies they are following and what are their existing practices so that with the help of this knowledgebase we can suggest refined methodology and can design training strategy to increase performance of teacher.

Chapter 3

New Trends in Agricultural Education

Agriculture is a state subject in the Indian Republic and hence, the fundamental responsibility for the manpower of this sector lies with the federal states especially Research System (NARS) with its vast network of State Agricultural Universities (SAUs) and ICAR (Indian council of Agricultural Research) institutes provides a blend of federal and central system..

Agriculture education in India may have its origin in the medieval period but organized agricultural courses were started in the beginning of the 20th century with the establishment of six agricultural colleges in different parts of the country. At the time of independence, there were 17 agricultural and veterinary colleges under the control of the state departments of agriculture and animal husbandry. Research and extension programmes were largely undertaken by the state agriculture and animal husbandry departments and to a very limited extent by these colleges. Overall the level of agriculture research and extension in the county was very low. This was in spite of the fact that certain research institutes were also established in the country which were operating independently at various places. The need was felt to reorganize the agricultural research system in order to increase food production through better and improved application of science and technology.

The Imperial council of Agricultural Research was set up on July 16, 1929 as a society, following Independence, it was renamed as Indian council of Agricultural Research (ICAR), As a result of joint Indo- American Teams of 1954 and 1959, reorganization of agricultural education was initiated by establishing the first agricultural university in 1960 at Pantnagar (U.P) on the Land grant pattern of USA. It was indeed a Watershed in Agricultural education in India. For the first time an integrated approach was adopted at Uttar Pradesh Agriculture University, Pantnagar in which by design agricultural education, research, and extension were allowed to prosper together. Presently there are 39 SAUs, deemed-to-be universities and one central Agricultural University in the country admitting 9500 students annually in UG and about 5500 students in PG Programmes. In addition, there are 46 private colleges admitting about 4000 students annually for agriculture related courses.

ICAR and SAUs are large employers of scientists/teachers/ extension scientists in the field of agricultural education, research and extension. The cadre strength of scientists in ICAR is just above 6000. There are about 19,000 scientists and academicians in the SAUs. At present, India has the second largest scientific manpower in the field of agricultural and allied sciences. However, certain facts have come to light. for instance, according to the World science report 1993, UNESCO, the total number of agricultural veterinary graduates in India registered a growth rate of 4.1 per cent and 3.4 per cent respectively and there were 1,62,800 graduates of agriculture and only 33,400 graduates of veterinary graduates in a country where livestock population is around 450 m and expected to grow to 650m. Even a conservative estimate of a total of 27,000 scientists in agricultural research and education, it works out to be 0.028 per 1000 population. This is indeed a very small figure even when compared with the figure of total scientists' per. 1000 population of 0.1 (world science Report, 1993, UNESCO) in India. There is certainly a strong case in point to promote more graduates in agricultural and allied fields considering the larger interest of the national goal.

Trends in Agriculture Education

The 20th century witnessed radical developments in science and technology. These developments had an impact on the social and economic changes that took place in the World. Agriculture related science and technology had major impact. This process of development was driven by some significant trends in the agricultural education. Some clearly visible trends that could be made out could be listed as under;

1. Agricultural Education as Agriculture and crop science.
2. Expansion in the domain of Agricultural by adopting various social science and advanced science discipline in to its fold.
3. Vocationalization of Agricultural Education.
4. Taking a comprehensive approach by inclusion of Agribusiness in the fold of Agricultural Education.

Agricultural Education as Agriculture and Crop Science (Till 1950s)

In the initial period major focus of agricultural education was on plant and animal science. This was primarily because of lack of sufficient scientific Knowledge in this field. The major concern at this point of time was to develop scientific know- how of agricultural science. At this stage lots of documentation and database development of prevalent agricultural practices took place, but all this was at a very initial stage.

Expansion in the Domain of Agricultural to Adopt Various Social Sciences and Advanced Science Disciplines into its Fold (1960s to mid 1970s)

The next important trend that could be identified was curriculum expansion, *i.e.*, inclusion of various other basis sciences and social sciences into the fold of agricultural education. At this point of time the priorities of government had changed. Now, it was not only developing agricultural as a science but to attain food sufficiency with its help. The primary area of focus was on developing high yielding and resistant varieties. Therefore, the disciplines like tissue culture, genetic engineering, rural sociology and agricultural economics started finding their in the curriculum.

Vocationalization of Agricultural Education (Mid 1970s to 1980s)

Once the food sufficiency had been achieved by the nation, the priority of the government changed to generating self- employment opportunities through agricultural education. Therefore, lots of vocational courses on commercial forestry, fisheries, dairy and horticulture etc. were developed and added to the fold of agricultural education.

In the phase Agricultural Universities had full- fledged college of Veterinary Sciences, Fisheries sciences, Horticulture and forestry. Various ICAR Institutions were opened up to specifically develop the knowledge in and serve these niche areas. These institutions and colleges started offering areas specific short- term as well as full fledged degree programmes.

Inclusion of Agribusiness in the Fold of Agricultural Education (Since 1990s)

It was time now to take a comprehensive look at agricultural sector. To assess its contribution to all the three sectors of economy. It no more remained confined to primary sector but spread its wings to secondary sector (*e.g.* value addition, food processing etc.) and tertiary sector (*e.g.* hotels and restaurants, ready to eat food segment etc.). it required further curriculum expansion and integration by including courses like Agribusiness, Food-science and Technology, clothing and Textile, and Export etc. other courses such as Entrepreneurial development, communication science, Behavioral Science, Marketing and Finance were also added to its fold.

Trends Needing Attention

In recent discussion among agricultural educators, six trends seem to come to the top of the list for intense study. These are critical issues worth attention.

Change in Dominant Culture

Our nation's population is becoming more diverse. How agricultural education adjusts to this diverse population may well determine its survival.

Urban Agriculture Growth

Agricultural Education in urban school is not new, at least in a few notable situations. However, the real growth potential for agricultural education surely rests in urban and suburban centers. The aggressive establishment of urban agriculture programs may be the most significant development for agricultural educators in the 21st century.

Curriculum Expansion

Some agricultural educators contend that the real staying power of the program can be attributed to the changes in the curriculum. The curriculum has changed to reflect new dimensions and career path in agriculture. A dynamic industry requires a dynamic curriculum. Is our curriculum a dynamic living curriculum?

Accountability for Program Impact

There is hardly any known institution that has not come under intense scrutiny regarding program accountability. The key questions are what is the impact of the program, what changed because of this program and what difference does it make?

Program Relevance

Relevance is critical to accountability and impact. Does the current set of practices and procedures in Agricultural Education have meaning to current and potential stakeholders? Are we struck on tradition or adjusting to meet relevant need of stakeholders and learners?

Upgrading Teacher Education

The key to learning is facilitation. Teachers are the ultimate facilitators. Are we preparing teachers in agricultural education in the best way or are we tied to an old model?

Obviously these and other trends could have profound impact on our programs especially if we take them seriously and get of the curve in shaping them. If we wait and react to them, it is too late. Although the trends identified here represent only a limited view of impending change, there is evidence that few things will never change. The bedrock principles upon which agricultural education operates are not going away:

- ☆ Experiential/problem-based Education
- ☆ Career Exploration in and about Agriculture
- ☆ Knowledge and skill in and about Agriculture
- ☆ Personal growth and Leadership Development
- ☆ Teamwork/cooperation and citizenship

Challenges and Issues in Agricultural Education

The agricultural education in India is faced with one of the biggest challenges. It has to identify its role in equipping the human resources for enhanced agricultural productivity and sustainable use of natural resources. Agricultural colleges and universities were initially assigned to disseminating scientific knowledge and skill to the farming community and to train the to use such skill for better output. As a backup for such a mission, agricultural research was encouraged to adapt the scientific knowledge to suit to the realities of rural societies. However, these initiatives could not keep pace with the fast changing scientific and technical improvements and gradually failed in their objective to cultivate the most modern skill and attitudes to both agricultural students and farmers. Developing a mechanism to update the curricula of agricultural education and extension. at least on a five-year regular basis, is crucial. This is very relevant to teaching, research and extension functions of SAUs as they form the inter-related, theoretical and practical basis of modern agricultural education in India. Moreover, the link between these three major strands is weak, particularly because extension is still not well recognized as the other two academic functions of the Agricultural University.

The Knowledge, generated by research, often gets obsolete by the time it is disseminated through education/teaching and finally trickles down to farmers and practitioners through extension due to big lapse of time. The Universities should consider this problem seriously and should reform their academic activities to strengthen the links between research, teaching and extension.

Economic liberalization and WTO regime demand manifold reorientations in our agricultural education and extension system. Market oriented agricultural education and extension along with changes in agricultural marketing policy for the national and international markets is the need of the hour. A cadre of social scientists with clear insight into agricultural related disciplines, such as agricultural economics, agri-business, marketing management, rural sociology, agricultural anthropology, agricultural ethics and politics, has to be built up to provide the back up for solving the problem of globalization in agriculture. This will imply total reorganization of agricultural education with emphasis on other agricultural related social sciences and effective agricultural policy research. Such e re-organization will empower our agri-graduates and scientists to face the challenge posed by demand of the national and international markets.

Any individual initiative needs certain motivation for sustenance. Pursuing agriculture as a profession/job for a considerable time is also influenced by motivational aspects such as monetary gains, employment, better quality of life and social acceptance. The number of agricultural producers is declining. The population that makes a living directly from agriculture continues to fall in developing countries. People, with rural background and engaged in traditional agriculture, are reluctant to let their children and younger generation to pursue agriculture as a full time profession. Those, with rural background and genuine interest in agriculture, do not have the basic education to go for higher education in agriculture.

This prompted rural youth to go in for non- agricultural courses and moving in of urban based agriculture graduates without any practical Knowledge of agriculture. Agricultural education has to be made more practical and rural realities oriented while agricultural extension courses should be tuned to serve the needs of the rural community through dynamic interaction with rural societies. Such extension activities would enrich the academic institutions and universities.

Developing countries lag behind in introducing educational courses on environmental and sustainable agriculture. In India, too there is a need to provide such a curricula re-orientation to academic institution to create an environment sensitive faculty and to help bring about attitudinal changes among rural communities.

Chapter 4

Higher Agricultural Education in India: Issues and Challenges

Agriculture is a way of life, and continues to provide livelihood to 650 million people. A rapid stride made in agriculture, changing from food scarcity to food surplus country, during last part of the century, was significantly contributed by development of effective human resource in agriculture, through setting up of State Agriculture Universities. The agriculture education, in past, was aimed to create Farm Managers and Extension Workers, needed for agriculture and specialists for research and education. The scenario of agriculture has changed, dimensions of problems are different than what was before. The strategies for agriculture development have different focus. The job opportunity, which was more in public domain, is now largely available in private sector. The jobs, which would have been taken up by the agriculture graduates, are being occupied by others having skills in changing scenario. What agriculture education needs is a matter of discussion. The human resource developed in agriculture education system has been instrumental for agricultural transformation in the country. However, the agricultural education system has not kept pace with the rapid technological development-taking place globally. During the early part of establishment of SAUs the academic programmes were structured to produce graduates who became primarily technology agents. In recent years, private sector has begun in big way to absorb some of our graduates. It is opportune time for many to become entrepreneurs In the present era of specialization and development in modern science, it is necessary that we restructure our agricultural education in a manner that the graduates coming out are analytical, technology oriented, sensitive to social and economic issues of farming in India, and able to meet challenges of new millennium with determination, zeal and commitment. The existing system of agricultural education is not fully equipped to enable agricultural graduates to capture other emerging job opportunities in the private sector.

Scenario of Agricultural Education in India

India has one of the most diversified agricultural education, research and development systems in the world. It is supported by about 30,000 scientific personnel including more than 7,000 of those engaged in active education, research and management. This vast network is coordinated by the Indian Council of Agricultural Research on the pattern of UGC as it does in case of traditional universities. In respect of NARS family, we are proud to have around 100 national institutions, 10 International centers, 5 National Institutes with deemed-to-be university status, one Central Agricultural University and 41 State Agricultural Universities. There are 558 KVKs and 8 Trainer's Training Centres for transfer of technology. This huge strength of human resource, infrastructure and resource network is the result of the continuous growth in priority sectors for effective education, research, extension and developmental activities in agriculture. In addition there exists a large network of large number of postgraduate and graduate institutions/colleges in the country that impart agricultural education. Several autonomous bodies including NGOs are also engaged in research and extension activities in agriculture and allied fields.

We have developed strong agricultural education system since independence. The establishment of State Agricultural Universities (SAUs) on the pattern of Land-Grant Colleges/Universities of the United State contributed to the reorganization and strengthening of agricultural education system in the country. The green revolution, with its impressive social and economic input would not have been possible without the contribution of these centres of learning in the form of development of trained scientific manpower, the generation and assessment of new technologies and their dissemination to ht farming community. The human resource developed in agricultural education system has been instrumental in agricultural transformation in the country. However, the agricultural education system has not kept pace with the rapid technological development taking place globally. During the early part of the establishment of SAUs the academic programmes were structured to produce graduates who became primarily technology agents in the present era of specialization and development in modern science, it is necessary that we restructure our agricultural education in a manner that the graduates coming out are analystical, technology oriented. sensitive to social and economic issues of farming in India, and able to meet the challenges of new millennium with determination, zeal and commitment. Majority of them should become job providers rather than job seekers. While the above steps are being taken to improve the human resource development of agricultural graduates, the glamour of government jobs dominates in our thinking, though in recent years, private sector has begun in big way to absorb some of our graduates to become entrepreneurs.

The future strategy for agriculture is going to be very different in view of increased population, increased demand for food and other commodities vis-à-vis over exploitation of natural resources, depleting land and water resources, and depleting genetic diversity. With the liberalization of economy, a number of multi-national companies with high-tech agriculture are entering into agriculture sector and the competitiveness is increasing. The challenge before us is how to reorient our agricultural education to make our graduates competitive nationally and internationally. Of late, there has been widespread consensus that quality of graduates in various disciplines of agricultural sciences does not match with the changing needs. Hence, there is urgent need for upgrading the quality and standards of higher agricultural education.

Challenges of Agricultural Education in Present Scenario

The agricultural scenario of the country in the twenty first century is radically different in view of the following facts:

1. The decline in productivity from 3.8 % to below 2% is viewed as a major concern of national agricultural research

system, which poses a serious challenge to effective management of natural resources and sustainable growth in agriculture.

2. Based on the experience of the last 50 years the free Extension Service has not proved cost effective as was envisaged. The farmers have been able to adopt only 30-35% of agricultural technologies in offer. The pace of adoption of new technologies has also been rather slow. Considering the present need of Indian agriculture it seems inevitable to introduce what can be called a Paid Extension Service that would invite private sector participation in our agricultural extension activities. It would also bring more graduates and technically trained manpower under self-employment.
3. Like extension, the research and education programmes in agriculture with the existing level of government funding will remain constrained in regard to facilities and resource mobilization. It will seriously jeopardize the quality of output. Hence, there is an utmost need to strengthen the linkage between SAUs and ICAR institutions and other organizations, and also to involve new institutions and organizations in the collaborative efforts to attain the common goals.
4. External factors like WTO, GATT, GATS, IPR, GMO and several other trades and export-import related policies have also contributed to the need of strengthening completive agricultural research and education system in the country. These factors also provoke to reorient the on-going research and extension programmes.
5. In order to compete successfully in the field of globalised agriculture, it is essential that agricultural research, education and extension system has a free access to information communication technologies and facilities so that its efficiency could be harnessed for the betterment of Indian farming community in this competitive era of agriculture.
6. The need for a change in the existing system is also visualized in view of the fact that more money and resources are needed for various programmes on account

of escalation in cost of equipment, infrastructure and salaries. The necessity for modern equipment and facilities in agricultural research and education system has long been felt. It is becoming apparent that limited resources of the government cannot meet the existing requirements. Hence resource generation has also become a primary focus of attention.

7. Distance education in agriculture has been perceived as an emerging area of intervention. In order to make it rewarding nonformal and vocational education packages will need to be developed. Collaborative efforts of SAUs, ICAR institutions, mass media organization and apex institutions like IGNOU and other open universities will, therefore, become inevitable.
8. Possibilities should be explored as to how SAUs can become global partners in imparting education in agriculture and allied fields.

The Context

Agriculture will continue to be a major contributor to the economies of the most developing countries. However, in many developing countries like India, the agricultural sector is undergoing rapid changes as a consequence of both technological progress and an industrialization process that calls for an increased market focus, competitive practices and higher productivity. Employment opportunities in off-farm agriculturally related activities are expected to increase at a faster rate than in production agriculture. Agricultural education curricula need to be redirected to address the changing nees and new challenges to usher in Green Revolution –II which will be different from the first Green Revolution in terms of approach and concept. While the first Green Revolution focused mailnly on food production to achieve self sufficiency, the Green Revolution-II will address not only more agricultural production from less land and less water, but also global competition in marketing, value addition in farm products, sustainable management of natural resources, environment protection and socio-economic issues.

Sustainable agriculture presents a deeper and more fundamental challenge than many researchers, extension personnel and policy makers previously assumed (Pretty, 1995). "Sustainable agriculture needs more than new technologies and practices. It needs agricultural professionals willing and able to learn from farmers; it needs supportive external institutions; it needs local groups and institutions capable of managing resources effectively; and above all it needs agricultural policies that support these features" (Smith, undated).

Everyone recognizes the critical role played by agricultural professional in transferring the new technology to end-users. They are involved in assessing farmers' technology needs, and technology development, evaluation and transfer. But there is growing concern whether today's agricultural professionals have adequate knowledge and skills to be effective in the current situation. "It is clear that if we as agriculturist are to make effective progress, we must change the way we plan, conduct, and communicate about research. Any component of a farming system can become the limiting factor to sustainability. It is therefore essential that those who work with farmers to develop sustainable systems are knowledgeable about the system with which they work (Reeves, 2000).

It follows that capacity building and professional development are fundamental pre-requisites for achieving the widespread adoption of sustainable agricultural practices. Particularly important targets for these efforts are those agricultural professionals that are so vital in bridging the 2-way farm technology gap the gap between what is known about sustainable agricultural practices and what is being applied at the farm level, and between what farmers know about sustainable agriculture and what researcher needs to learn. "Much information is unavailable or in accessible, particularly to poor farmers, many practical lessons have been learnt but not shared, and there are few opportunities for dialogue to enable concerns to be resolved". (FAO, 2000).

The universities have to develop and strengthen for both formal and non-formal education programmes to bring academic excellence relevant to future needs. Though, the curricula have been modified, physical facilities, equipment and teaching aids have been updated to some extent and facilities have been made for training of teachers in educational technology at National Academy of Agricultural Research and Management (NAARM), Hyderabad and Academy of

Agricultural Research and Management (AARM) in some SAUs, they are far from adequate. Teachers in SAUs are selected mainly on the basis of their professional degree and experience. The problem of inbreeding in the faculty in SAUs has assumed an alarming proportion. Further more, the beginner teachers do not have any formal pedagogical preparation. Training of faculty was greatly aided during the decades of sixties through various faculty developments and exchange programmes auspices of US Technical Co-operation Mission and the later the USAID Six American land grant universities participated in the faculty exchange programme with eight SAUs in India. Similar arrangements under Indo-US Knowledge Initiative on anvil.

Educational systems are dynamic in nature, continuously adapting to meet the emerging needs of society. In the era of WTO it is very much necessary to improve the quality of teaching for professional development of teachers as well as students. More than ever before in our history, education will make the difference between those who will prosper in the new economy and those who will be left behind. Thus, it is necessary that improvement programmes are run to improve quality of teaching sooner than later. Professional development of agricultural education demands revolution in teaching.

Teaching Agriculture

Agricultural Universities have adopted Practical Crop Production (PCP), Rural Agricultural Work Experience (RAWE) and Agri-Business Management (ABM) which have a deep impact on quality of education. Education in Agriculture needs a different treatment which should be experiential in nature. Presently, there is urgent need to move away from didactic teaching and look for new instructional approaches like student-cantered teaching, problem solving, action learning, experiential learning, project mode, etc. It is necessary to experiment with methods of teaching for improvement of quality and professionalism. An educational system is a group of components like student, teachers, media, materials and educational objectives working together in organized fashion. Application of innovations of science and technology along with humanities, social sciences and education is mandatory in achieving the goals of education. It is here that science of educational technologies comes in picture to analyse the process of education and solve the problems of teaching and learning. Thus, advances in educational technologies must now be used to systematically design, implement and evaluate educational process. It is necessary to experiment with methods of teaching for improvement of quality and professionalism.

Linkage for Reorganization of Agricultural Education:

Agricultural education at present is being implemented at four levels:

1. Doctoral level
2. Master's level
3. Undergraduate level
4. Vocational

While the country is proud of having four deemed-to-be-universities of national importance supported by ICAR for higher research, it would be desirable that more emphasis is placed on doctoral education at these institutions and only on master's level programmes in subjects which are not available in SAUs. This would facilitate better utilization of resources for higher education. SAUs generally cater to the regional/local needs in their teaching and research programmes which are effected through doctoral, master's and undergraduate programmes. Its main emphasis, however, has to be on improving the quality of education by making course content and curricula need based, more practical and field oriented without compromising on theoretical knowledge. Periodic updating/review of syllabi will be integral component of quality-oriented education. A quality based teaching/research programme will make the graduates and postgraduates of SAUs better scientist, extension workers and entrepreneurs in future.

Linkage for Strengthening RAWE

The programme of Rural Agriculture Work Experience (RAWE) was initiated in all the SAUs to build up the confidence of the students in the application of agricultural technologies in rural setting and in other management as well as operational aspects. For this purpose students are required to have multi-institutional experience and exposure through learning by doing. Hence, linkage with ZRS, Zonal ICAR Centres and other institutions including KVKs and agro-based industries would be essential for all the SAUs.

Utilization of Facilities at Local Institution for Postgraduate Education

To meet the specific training requirements at master's level it has been experienced that the requisite facilities like fields, equipment, funds, etc., are sometimes not available at postgraduate colleges and SAUs. On the other hand such facilities are available at zonal research stations of the SAUs and also at local research stations of the ICAR and CSIR. With a participatory spirit ensuring coordinated efforts for improvement in education, the graduate and postgraduate students can be attached to these stations for having a better training and optimum utilization of facilities at the local research stations. This will also help in strengthening the activities of the local research stations by adding manpower on research programmes without additional investment.

Constitution of a Coordination Committee for Research and Extension

Each agro-climatic zone is represented by SAUs, ICAR institutions and other local research stations/sub-centres. There also exist other institutions like traditional universities, CSIR institutes and DRDO laboratories which are engaged in similar research programmes. There is no coordination on zonal basis among these institutions to make research programmes of these institutions inter-dependent and supportive of each other. Coordination among these institutions will facilitate initiation of new areas of research thrusts. Unfortunately, the location specific requirement of education, research and extension does not remain the jurisdiction of only one institution as at present especially in view of the fact that for the same role we have several institutions in each agro-climatic region. Hence, effective coordination is essential to avoid duplication and multiplication of similar research programmes, facilities and equipment. It will also make the joint efforts cost-effective in attaining the goals of education, research, extension and transfer of technology. In view of this each agro-climatic zone may have a coordination committee headed by Vice-Chancellor of the SAU representing the specific agro-climatic zone and heads of the different ICAR and other institutions as members. In many cases in the same agro-climatic region there may be two SAUs. In that case the two Vice-Chancellors could act as Chairman on rotation basis. However, considering the apex role of the deemed universities they may have a separate coordination committee headed by its Director as chairman and other heads of the institutions including Vice-Chancellors of the SAUs as members. It is hoped that such coordination committee will be able to set the agenda based on the national/regional requirements to provide supportive role to the apex institution. This will also solve the problem of technology transfer by those institutions where requisite staff and facilities have not been provided by their respective departments.

Linkage and Coordination for Training and HRD

In view of the fact that the SAUs are multi-faculty organizations established on the basis of the location specific needs of the country, these are obliged to focus on the HRD through its various degree programmes. At the same time some of the programmes of human resource development through training and refresher courses as undertaken by ICAR need to be decentralized from national level to SAU level where the research scientists, teachers, trainers and extension workers can be provided additional facilities and orientation and refresher courses on periodical basis. The trainer's training on regional basis through SAUs has a greater relevance than providing the same through national level institutions. However, in frontier areas of research and education we may continue the system of having orientation and refresher courses through national/apex institutions. In such highly specialized areas ICAR may itself arrange trainers' training.

In order to make higher education/research more rewarding time has come to evolve appropriate strategies for developing sandwich Course programmes involving SAUs and ICAR institutions. These programmes would envisage participation of various institutions jointly to undertake teaching and research programmes of students pursuing higher studies The choice of participating institutions in this regard would depend on the specific requirements of the students/region.

Possibility should also be explored to make reciprocal arrangements for higher education especially at the centres of excellence among SAUs. Such an arrangement would permit sharing of seats especially at doctoral level in the identified areas of study by two or more SAUs and/or ICAR institutions.

Emphasis on Sharing of Research Facilities

The research requirements of different institutions cannot be tailor-made. Hence, they need to have mandated based facilities and resources. It is experienced that all the institutions do not have requisite research facilities and resources. Hence, sharing of facilities has been realized as one of the options for meeting such requirements with little or no investment. The national institutions possess the facilities and equipment which cannot be multiplied due to its high costs. Hence scientists of SAUs and other national institutions may share such equipment and facilities. Similarly at SAUs, there are many facilities which are not available at the sub-centres of ICAR and zonal research stations. These centers may be provided necessary facilities on sharing basis from SAUs. While doing so it would be imperative to identify the distinct equipment and facilities to be shared by the institutions. Suggestions given below may also be considered for effective coordination among various institutions working in the state:

1. In many states, ICAR and CSIR institutions, and SAUs have been established for carrying out teaching, research and extension activities. The coordination among these institutions is negligible or non-existent. There is a need to strengthen the linkage among the institutions to bring out synergies in implementation of common research and extension agenda.
2. A number of research projects are being carried out almost on similar lines by various regional and central institutions resulting in substantial mis-utilization of resources. This problem could be minimized by networking of common research programmes of the institutions.
3. At present, there is a disparity of pay package of the supporting research staff particularly JRF/SRF/PDF, working in SAUs and ICAR institutions. The research staff working in SAUs is being paid a lower salary than their counterparts in the ICAR institutions having similar qualifications. There is a need to sort out such discrepancies among the institutions.
4. There is an utmost need to promote interaction among students/trainers of SAUs and ICAR institutes in co-curricular and extra-curricular activities on periodic basis.

5. Counselling and other activities in SAUs may be funded by ICAR with a view to strengthen these programmes and development of human resource.

Coordination Between CSIR, DRDO and Traditional Universities

In each agro-climatic region we have a number of research centers belonging to these organizations having mandate. The coordination in terms of research and education and, wherever possible, for transfer of technology will ensure attainment of goals and efficient utilization of resources on mutual basis. The special areas like biotechnology, microbiology, biochemistry, molecular biology, agrometeorology, environmental sciences, satellite and surveillance services and others can be taken up jointly by SAUs, ICAR, conventional universities and other autonomous organizations which have been established in specific agro-climatic region.

The aforesaid areas provide avenues for ensuring normative, operative and enabling linkage. It also ensures attainment of goals with existing resources and capacities for agricultural growth at national and international level.

Sharing of Communication Network for Distance Learning Programme

The ICAR has been given a new mandate to provide education for school dropouts, neo-literates and practicing farmers, farmwomen, fishermen, etc., through distance learning mode. The SAUs and ICAR institutions can take up such programmes in collaboration with broadcast organizations and open universities including IGNOU. The information support through print media, radio, TV, Video-Conferencing, IT Kiosks and counselling at Study Centres could be taken up jointly so that lessons for vocational programmes are prepared and delivered, and students are evaluated on the basis of their performance in theoretical as well as practical aspects. In this task there is a need for central assistance through ICAR or other agencies to have FM radio stations with each SAU and necessary support of the Ministry of Information and Broadcasting to ensure access to knowledge through existing mass media network for distance education.

Linkage for Strengthening Agricultural Extension

The SAUs are now considered as the resource institution for generating technologies in various fields of agriculture and allied subjects. The ICAR, DRDO and CSIR institutes have also remandated their programmes by inculcating Transfer of Technology as the major component in their programmes. At present the Transfer of Technology may be envisaged through field extension work and farm advisory services, technology testing and refinement, training and orientation of functionaries, training of farmers and vocational education, production of information materials for distance education, organizing Kisan Melas and many other extension programmes for benefit at large. All these programmes require coordinated efforts of different institutions including NGOs and line departments of different agro-climatic regions. Considering their involvement in these programmes following linkages are suggested:

1. While SAUs have been designated for nodal responsibility in respect of transfer technology through different KVKs and line departments it will also be relevant to suggest participation of the scientists representing ZRC and other local research stations to work as resource persons in the training programmes and also in the development of information material. If national institutions are organizing HRD and other training programmes the scientists from SAUs should participate as resource persons in the relevant areas.
2. At present in addition to KVKs, extension work is undertaken through line departments and also through SAUs, private input agencies and NGOs. This role has to continue at all the KVKs and other institutions.
3. Coordination of Extension Programmes on Agro-Eco System basis, *i.e.*, all SAUs, ICAR Institutes, CSIR, DRDOs, ZRS should have uniform extension strategies for different micro farming situations.
4. Each KVK has Scientific Advisory Committee represented by host institution, ICAR institute, district line department and people's representatives. It may also have representatives of other institutions like CSIR, DRDO and NGOs for having wider scope of work.

5. The distance education has emerged as a potential means for non-formal education, which could be made effective by production of appropriate instructional material depending on the local needs, problems, resources availability and other socio-economic conditions. Production of information material for distance education requires a coordinated effort of specialists in communication, educational technology and those representing the subject matter areas like agriculture, animal husbandry, fisheries, horticulture and forestry with responsibility for production of total information package and counseling as we have at study centers in other programmes.
6. In each agro-climatic region there is a network of mass media like Radio, TV and local newspapers. An effective linkage has been found to be helpful in rapid dissemination of information and technology among the farming community.
7. Privatization in extension has emerged as a novel area on account of change in global trade and export policy in agriculture. The Paid Extension System has a major limitation as it leaves those people out who cannot afford to get the advice without payment. On the other hand those who need it, can be benefited immensely through this system. The new system is also amicable to provide more jobs to the graduates and technicians. The SAUs and specialized research stations may provide technical back up through periodic training and orientation for the persons who are engaged in private extension service.
8. In developing an appropriate mechanism for collaborative action it may also be necessary to involve line departments and other agencies for rural development. The SAUs have extension advisory committees for similar role. However, the transfer of technology is the mandate of other research institutions also for which the needed support may be provided by the SAU scientists to the neighbouring institutions on mutual basis.
9. At present the Zonal Coordination Units dealing with Transfer of Technology have been established on the basis

of convenience of the zonal coordinators and the availability of land and other facilities. However, the basic issue for having similar programme on agro-eco system basis has been ignored thereby two or three types of agro-climatic zones are represented in most of the zonal coordination units. This has to be avoided through reorganization of the area of jurisdiction of zonal coordination units.

10. Globalization and privatization in agriculture has provided the opportunity for initiating private extension service in agri-business programmes. The demand for agricultural graduates and technicians is likely to grow with a change in the nature of extension system. While the SAUs are presently engaged in producing graduates and postgraduates. At the same time, there is need for training of technicians and the same is yet to be taken up by SAU-ICAR system. The KVKs have the requisite infrastructure to provide need based vocational training in agriculture and allied fields ranging from six months to one year duration. These trained technicians have potential for self-employment and also for rendering services to the farmers, entrepreneurs and private extension system. The sub-centres of SAU and ICAR may also take up this type vocational education in specialized disciplines depending upon the facilities available with these institutions. For organizing such courses it may be appropriate to design and develop the curriculum by involving the scientists of SAUs as well as ICAR institutions so that these model courses may be implemented elsewhere also.

Linkage in Recruitment and Examination

In order to ensure that quality human resource is allowed to work at SAUs, it is desirable that a uniform recruitment policy is adopted in SAUs as well as ICAR institutions. Similarly, all SAUs, and if possible Deemed Universities, may also have a common entrance examination for the selection of students at Graduate, Postgraduate and research levels. This will ensure uniform quality of admitted students in SAUs and Deemed Universities. The provisions can be extended to the placement services of graduates/

postgraduates in order to ensure best possible utilization of human resource.

Quality Assurance in Agricultural Education

A holistic approach involving SAUs, ICAR and Deemed Universities needs to be adopted to ensure quality of institutional outputs in the form of research information and new graduates/ researchers. Besides ICAR, funding for integrated programmes can be generated through appropriate industry-institution linkage. A viable mechanism for ensuring quality should be developed jointly by SAUs and ICAR institutions including Deemed Universities. It will be worthwhile to consider the feasibility of merging old/less productive research units of ICAR with the regional SAUs.

Some other measures than can help promote quality education include enhancement in the number of National Talent Search Scholarships and contingency grants to the students from Rs. 1000/ - to Rs. 1500/-. New programmes of education should be initiated only after the approval of ICAR and Accreditation Board. Compulsory courses in Instrumentation and Computer Application should be included in all postgraduate programmes. New colleges within and outside agricultural university system should be established after going through accreditation procedure.

Linkage for Strengthening Library Services

Due to rapid progress and adoption of information communication technology in the field of library sciences it has become easier to ensure access of students and faculty members to the latest books and journals from every corner of the world. The information technology has also provided facility for translation of the original versions to facilitate readers in the languages convenient to them. While every university cannot afford to go for sharing library facilities at global level it may be advocated that the storage and retrieval of information may be done through apex libraries like IARI, IVRI, National Informatics Services and other institutions identified for the purpose. From these libraries information and material may be shared by SAUs, ZRSs and other agricultural institutions. This will save time and energy in management and safety of the books to be maintained in various libraries. The MANAGE has already started 'Library Without Books'. Similar facilities may be created elsewhere also for making education and research more pragmatic and effective in the Indian context where access to costly books and periodicals is becoming increasingly difficult for the students.

Updating Curriculum

There is need to continuously revise curriculum according to the needs of the market. Paricipatory curriculum design should be practiced to gauze market and bring consensus among stakeholders. The curriculum must focus on development of employable skills, and strategies should be designed to develop those skills. This calls for innovations in teaching methodology. The following subjects need to be included in the curriculum of agriculture graduates:

1. Good agricultural practices for field crops, fruits and vegetables
2. Biotechnological tools for crop improvement
3. Organic farming
4. Micro irrigation and fertigation
5. Agribusiness
6. International marketing, WTO and IPR
7. High-tech agriculture and precision agriculture
8. Value addition and post-harvest processing
9. Expert system and ICT applications
10. Hands-on training in agri-entrepreneurship such as quality seed producton, mushroom production, apiculture, commercial horticulture, floriculture, bio-fertilizer production, biological methods of plant protection, food processing.

The success of academic institutions in agriculture will be known by their responsiveness to change. Significant social forces have created both opportunities and challenges. The need of the hour is to capitalize on the strength of our knowledge and experience and manage them creatively to add value to the products. Are we ready with needed attitudes and skills to achieve success?

Soft Skills Needed

The students in agriculture require competencies in group discussion, panel discussion, analytical ability, which have received less attention. They also require confidence and techniques to deal with the situation,. To inculcate the foft skill the faculty with above skills and experience would be needed. Coordinating seminars, workshops, panel discussions, organizing inter-institute competitions, conducting research, conducting trainings, supervising projects, developing case studies are some of the approaches to improve the soft skill of the faculty, who can provide the skill to the students.

Education, a means of empowerment and social development assumes much significance, when agriculture is added to it, since agriculture is looked upon for employment with multiplier effects and economic development.

Student's Activities at Zonal and National Level

While sharing of physical and financial facilities is considered an important means of collaboration, sharing of heart and mind of the students and faculty members through inter-institutional programmes like sports, quizzes, debates, essay competition and cultural programmes is equally important. Activities in this regard could be organized by SAUs and ICAR units on a regular basis.

In Search of Excellence

Preparations are being mode for moving towards excellence in the frontier areas of science, for converting the scientific knowledge into wealth for the economic progress of the nation and for reducing the sufferings of the poor people. Alleviation of poverty can only be done by providing employment opportunities to the poor people, meeting their nutritional and health requirements. Indian agriculture has changed significantly during the second half of 20thCentury and after a lull, the shift is gathering momentum in preparing the farm sector for not only into the 21st Century but also into 3rd millennium A. D. as a modern and strong sector with an unambiguous identity in international arena. Farming which in India was taken as a means to provide food and fuel to the family suddenly turned into a profit earning entity. From mere subsistence to profit earning vocation, farm sector in India is poised to turn into a dynamic commercial player ready to play its role in global playgrounds. From "learning to know" through learning to do "and learning to be", the next millennium will belong to' learning to live' Education in today's world and more true for the world of 21 st Century has become means of living, a source of survival for the mankind. Strange is the world today, Not the nuclear weapons, but front line technology generated through " The cutting Edge Research" has become the most powerful weapon and most precious wealth which gauges the strength of a nation. A nation's accomplishments could well be judged by the standards and the degree of excellence that its universities and research institutions have achieved. At least some of the developing countries like India, China, Korea, Taiwan and few others in Asia and elsewhere are trying to compete with some of the top countries in the world to push forward the frontiers of knowledge and creating more efficient technologies in order to strengthen their economics. While doing so, unless the quality of scientific and technical education and place antisocial status of the teachers in society is raised to an appropriate level, this will be futile to dream for it. This essentially is now or never case and an immediate serious thinking is required at all levels on this important issue. The future educational requirements shall be highly specialized and sophisticated and creation of nothing less than excellence in every field shall be needed. Improved knowledge thus generated will have to be disseminated and taken to the actual user, may it be the farmer or the agro-industrialist. Industry is going to be an important player in rapid transformation of farmer and farming scenario of India in the first quarter of the 21st Century.The huge investment that agricultural education and research will require, can not be borne entirely by the state. Industries will have to come forward to share it since it is going to be a beneficiary. Industry as a matter of fact will be playing the role of a major participant. Graduates of today will be meeting the challenge as they shall be the human resource needed in the entire process.

Points for Reorientation

Education, a means of empowerment and social development, assumes much significance, when agriculture is added to it. Efforts need to be made made for making agriculture education more relevant by modifying course curriculum, enhancing learning process, imparting values through personality development and making teaching an active process. There is a great quest allover the world for the synthesis of knowledge and culture. Ancient knowledge is being rediscovered in the context of modern knowledge. The humanists and technologists are finding themselves in greater and greater need of each other. The educational implications of these reflect the latest trend of synthesis and our educational objectives must improve the idea for preparing new kind of men, who can consistently and progressively harmonize within himself the broad vision of humanist and skills of the technologies. Thus, agriculture education has to add values, to be relevant, with development of skills that are needed with changing time and the career planning has to keep pace, which must improve in the current context. Development of the ability to apply knowledge for solving problems, analytical and diagnostic ability for decision making, development of imaginative and innovative solution to deal with the situation, leadership and team spirit in handling the problems, 'entrepreneurial skills and motivations are essentially required to face the challenges in agriculture. The students must have the knowledge of Information Technology, communication skill, leadership qualities-engineering of agriculture education system calls the tomorrow's agricultural graduate. Each education system must have its own vision of its products within the context of their internal and global requirements for development.

Future Outlook

The agricultural graduates of tomorrow should posses respect towards his or her culture and tradinational values, practical knowledge of field problems, a sense of equity and social justice and selectivity in approach compatible with contemporary social, cultural, ecological and economic conditions. He should be industrious. He is resourceful and cleaver enough to deal moderately with risks and difficulties, creative in pooling together. Indigenous and other resources into new quality products or ways of doing things with a strong sense of workmanship. He is a team player for while he is working individually in his own enterprises, he can also work harmoniously with others towards the fulfillment of the groups goals. He should be thinking of others, especially those working for him of the opportunities for advancement he can provide or their personal welfare and development. Networking would be facilitated and standards of quality adopted towards global collaboration rather than global completion. Can agriculture system be modernized with focus on small farm technologies in an era globalization and the demand for the economy of scales? How fast can the education system in agriculture re-engineer its deeply-rooted academic producers into flexible curricula developed with the participation of industry; and for industry to restructure its hiring systems to give better incentive to blue collar occupations? Can the present constancy-based agriculture education programmes still be relevant in emerging knowledge-based economics, information technologies and borderless thinking and learning revolutions? It still is, how can it be made relevant? The students in agriculture require competencies in group discussion, panel discussion, analytical ability and confidence and techniques to deal with the situation. These value additions are becoming relevant and require serious consideration.

Chapter 5

ICT Applications in Agricultural Universities

Introduction

Information and Communication technology, commonly abbreviated as ICT, is the catch-all phrase used to describe a range of technologies for collecting, storing, retrieving, processing, analyzing and transmitting large volumes of information, hitherto unthinkable. These computer centric technologies of recent origin are advancing at an unimaginable pace. Its impact on all facets of human civilization is stupendous and has never been felt before due to any technology in such a small time.

Advances in ICT have progressively reduced the costs of managing information, enabled individuals and organizations to undertake information related tasks much more efficiently, and to introduce innovations in organizational structures, processes and products.

There is bound to be a complete change in the realm of education under the impact of ICT. Access to information is a few clicks away, thanks to ICT. Information, thus, is no longer a preserve of a privileged few. Knowledge, as the saying goes, is power. So, ICT has the potential to enable a person to reach his or her private zeniths by offering him or her volume of information available in the public domain.

No modern institution of higher learning can sustain without adopting information and communication technology in all facets of its activities – be it teaching, research, extension or management. Implementation of ICT enabled activities will help the institutions to achieve their goals more efficiently and speedily.

But, that will require a clear understanding on the part of institutions as to what ICT applications can do for them in perusing their strategic goals. Moreover, one should appreciate that implementation of ICT cannot take place overnight because necessary financial resources may not be available at one point of time. At the same time we have to accept the reality that there is an acute shortage of trained and experienced personnel in the area of ICT in all universities, who can spearhead the implementation. Assimilation of ICT into the university operations must deal simultaneously with the realities of its organizational culture, general planning culture, existing technologies and systems and relevance of ICT in the general long term goal of the university.

All these point towards the fact that a university should have a policy framework within which ICT enabled activities are to be carried out in a phased manner. It is to be borne in mind that mere introduction of a new technology will not be sufficient. What is required is a complete change in thinking processes of all concerned to assimilate ICT applications in their working habits. Unless that is done we shall not able to sustain the gain accrued from technical success.

The present paper addresses the following three issues pertaining to implementation of ICT enabled activities in an agricultural university:

- ☆ General vision of the role of ICT to meet the long term goal of an agricultural university.
- ☆ Planning and development of ICT infrastructure.
- ☆ Development of manpower in the area of ICT and enrichment of ICT skills in the university on a perpetual basis.

Role of ICT

The three basic pillars of Agricultural universities are teaching, research and extension services. The role of ICT in each of these activities cannot be belittled. Apart from this an efficient and transparent management of the universities can be ensured by using ICT based management systems. It is imperative that the universities develop basic ICT infrastructure to realize the advantages of application of the technology by progressively bringing more and more of their activities under the ambit of ICT.

ICT in Teaching

Throughout the world higher education is slowly and steadily undergoing a paradigm shift from an instruction-centric university model to learner-centric integrated network model based on access to learning resources and students' initiative. The rate of change is conditioned primarily by socio-economic factors and by the availability of technologies.

An important fallout of ICT revolution is that group learning is being progressively replaced by individualistic learning. The latter encourages a democratic learning process in the sense that the learner can choose what he wants to learn. But, it appears that a disciplined learning will still require conventional class room teaching for time to come. What is required at this stage of evolution is a synthesis of conventional class room system and new technologies. The latter provide effective tools that complement conventional teaching practices. Audio visual aids make class room teaching more explicit, interactive and perhaps more interesting. Multimedia packages facilitate and encourage individualized learning. Provision of internet connectivity in the class room makes the whole world shrink in the small class room. We shall have 'smart' class rooms as opposed to 'passive' class rooms.

Information technology has helped in changing the content of programs and curricula by making new information easily available. The regular updating of courses is now possible without waiting for publications of books incorporating latest information, which takes a long time. By the time a book is published many of the information contained in the title may have become obsolete. The internet brings the current information on any subject matter at the desktop at the asking.

Intranet communication inside the campus will help to have more meaningful interaction between a teacher and his students beyond the class room. It will be possible for the teacher to pay attention to individual student which may not be possible in a non-monolithic class.

Deployment of ICT in teaching enables a qualitative change in the delivery of an educational system due to the reasons outlined below:

- It encourages innovative teaching methodologies and more interaction with students.
- It enables better management of class rooms and students.
- It helps the teacher to remain a perpetual student.
- It allows optimum utilization of resources and facilitates sharing of resources with other institutions.

However, the creation of smart class rooms that form the backbone of the new teaching methodology is only the first tentative step. It would also require an updated knowledge of teaching the subject and teaching methods on the part of the faculty. The biggest challenge to the effective utilization of ICT in class room teaching is a proper teachers' training in ICT environment. Teachers may resist any change in their traditional roles and may refuse to adapt to the new situation.

ICT in Research

Today's research in the field of agriculture is focused towards a deep understanding of ideas, processes and practices. It touches or covers many disciplines and therefore, requires knowledge in diverse areas on the part of scientists. The scope of information has become wider in present day research as it is now available not only in print materials but also in non-print electronic forms. A fast retrieval of the desired information is the hallmark of ICT. Navigational tools and filtering mechanisms are provided in World Wide Web search sites. These are of great utility to scholars and research workers to find documents on a very specific subject. The role of reference librarians providing these services in a traditional way is all but over.

Processing, analyzing and presentation of research data have become fast, accurate and meaningful due to readily available software. Life has never been as smooth for researchers as it is now. Research outputs are bound to increase. Another aspect of ICT is Expert systems. The innovative ideas and practices which are in the domain of tacit knowledge can now be captured in Expert systems. This gives commercial leverage to universities and the research workers.

ICT in Extension Services

Through the extensive use of internet connectivity and agricultural portal it will be possible for the universities to reach to the rural population in an effective and economic manner. Research findings quickly pass on to the users in the field. The KVKs are better managed and their impact on rural areas will be better felt. Creation of university's web site goes a long way to achieve this goal apart from a secured and fast connectivity.

Digital Library

A digital library is one in which learning resources are available in machine- readable format accessible by means of computers. The digital content may be locally held or accessed remotely via computer networks.

Compared to traditional libraries which are limited by storage space, digital libraries have the potential to store much information simply because digital information requires very little physical space to contain it. The cost of maintaining digital library is much lower than that of a traditional library. Further, digital libraries can immediately adopt innovations in technology. It also provides access to much richer content in a more structured manner, i.e we can easily move from catalog to the particular book and then to a particular chapter and so on. However one has to be cautious about copyright law and access is to be restricted to registered users.

Most digital libraries provide a search interface which allows resources to be found. Some libraries create special pages or sitemaps to allow search engines to find all the resources.

An attempt to create a digital library is the first step an educational institution has to take for implementation of e-learning.

ICT in Distance Education

The status of development of a country largely depends on level of education imparted to its people. Developing nations desperately need and constantly seek an educational system that equalizes opportunities for the poor, the disadvantaged and the women in general and for those living in rural and remote areas in particular. The Agricultural universities are to realize this historical responsibility in the ICT era. Access to good educational programs can be greatly extended to large audiences in rural and remote areas through the use of new communication technologies. The cost effectiveness has been proved beyond doubt.

Having established a digital library to be accessed by individuals in the campus, the university should take up web-based distance education program in specific areas of demand. This will call for activities in two directions. Firstly a knowledge environment for web based learning is to be created employing available technology. Secondly the curricula, reading material, on line evaluation and monitoring processes are to be standardized. Web-based distance learning system will require continuous training of staff involved to cope up with the continuous technological change.

Office Computing Services

Office computing is used here for the application of desk-top computers to support general office tasks. These include word processing, document storage and retrieval, access to internet services, use of intranet services for accessing selective university data bases, store information formatted as web pages etc. It does not only apply to clerical and secretarial work, but also to the office work of managers, researchers and teachers. To be effective it will require appropriate computer systems and software on the workers' desks. E-mail and access-to-internet services require specific application software which is currently integrated in all modern operating systems. Effective office automation will also require adequate knowledge and skills in the use of computer systems and different office applications.

Information Systems

Management Information system (MIS) refers broadly to a computer-based system that provides managers with the tools for organizing, evaluating and efficiently running their systems. MIS includes software that provides databases and helps in decision making. Necessary hardware and network support will be required for users at different geographical locations to access the system.

In any organization a number of distinctly identifiable management activities are carried out. Each of these activities needs a separate MIS operating on specific database(s). Most of the databases are dynamic in the sense that these are updated regularly. One information system may link another system for retrieving a part of the database of the latter.

A few important information systems to be developed by Agricultural Universities to have an efficient and transparent management are:

1. Academic information system
2. Financial information system
3. Stores and Procurement information system
4. Library information system
5. Research information system
6. Extension information system
7. Human resource information system
8. Archive management information system

None of the systems can be implemented overnight due to the following reasons: (*i*) a lot of time is required to procure software and necessary hardware; (*ii*) necessary resources may not be available at a time; (*iii*) network infrastructure may not be available between users. So, the implementation needs to be carried out in phases.

The first stage of the implementation is the documentation of the detailed functional requirement and necessary database for each of the systems to be developed. The actual implementation of the system, then, may be outsourced. Irrespective at which phase and time a particular system is implemented, the pre-implementation functional study document in respect of each of the information systems needs to be prepared at the university level by utilizing available resources.

ICT Vision

It will now be prudent to spell out what the universities should envisage to achieve over time, with the support of ICT. Broadly speaking these are:

A communication channel that will ensure quick flow of digital data inside the campus covering faculty buildings, hostels, administrative units, library etc and between the campus and the outside world using internet connectivity of proper bandwidth.

- A digital learning system (including teaching, research, extension and distance learning) with a decentralized library system at the core.
- A website of the university to be updated on regular basis.
- A virtual paperless administrative system.

The long range ICT objectives need to be broken down into short range goals identifying the actual resources required so that resources are arranged in time and used as efficiently as possible. The success of any ICT policy is dependent on the approval and support from top management whose commitment will foster a climate in which a desire for innovation through ICT percolates down the line.

Successful development, implementation and deployment of ICT services depend on careful and realistic planning. However, planning as such cannot guarantee success. The risk inherent in today's ICT environment in any university is a general inertia (particularly at the senior level) to new technology which changes at a fast pace, increasing user aspirations (particularly among the new entrants and students) and tendency to view ICT services as a ready made product supplied by a service provider. Every individual drawn from the top management, teachers, and research scholars and down to the lowest rung of employees has to be an active participant in the successful implementation and maintenance of the policy. This will call for perpetual training programs ('I teach you and you teach me') of varied levels run in the university. It will also require that a few core level managers keep themselves apprised of technology and application trends. The users' responsibilities cannot be belittled.

End Users' Skill Development

An ICT environment ensures that academic, administrative and managerial processes are automated. This demands from the users the necessary skills to utilize the services/systems, keep them running, and develop them.

The university has to ensure that all students, academic staff, administrative and support staff, and managerial staff are trained on a continuing basis so as to equip them with the requisite skills to fully exploit the ICT environment in their different functions

End users' skills development includes all efforts to inculcate awareness, and general and specific computer skills related to the use of information technology. An end user is that person who uses ICT services or the information available from these services to support his/her learning, teaching and research, administrative, secretarial, or managerial tasks. As of today, the knowledge and skills for the development, implementation and use of ICT services/systems are miserably limited in both academics and administration. Consequently staff training assumes a major role in the successful implementation, use and growth of ICT in the university.

Permanent intensive staff development center needs to be set up as soon as possible. The contents, level and target groups of a training session need to be identified in line with actual requirements.

There should be a consistent set of training programs with different levels for different categories of (potential) ICT users. The long range goals of the training programs should ensure that end users:

- Use ICT services/systems effectively and as independently as possible.
- Be aware of the shared responsibilities for equipment, software and data, and create an atmosphere of collective responsibility.
- Contribute to the specification, design and implementation of IT applications.
- Sustain effective, efficient application and data management and system maintenance.
- Manage complex processes like ICT networks and systems.

The short and medium term goals shall be aimed at creating as rapidly as possible a sizable proportion of the staff who are familiar with, and able to effectively use ICT infrastructure in their daily work,. This may be viewed as the first phase to be followed by more advanced phases. After the first phase the university should expect that:

- All students and staff at all levels are able to use standard application packages, e-mail and internet

- ☆ Administrative chores like calling meetings and distribution of minutes and other documents are handled via e-mail.
- ☆ Staff-student interaction outside the class room should rely more on communication through e-mail and online message boards.

In order to achieve an effective training the university needs well-equipped training facilities. It is also necessary to develop on line training support materials like online manuals, frequently asked question list and answers.

Students' Participation

No ICT environment in an institution of higher learning can be effective without the active participation of students. Senior students having adequate competence are to be involved in creation of ICT environment. Depending upon the level of competence, students are to be involved in (*i*) training programs, (*ii*) website updating, (*iii*) functional requirement studies, and (*iv*) digital library processing activities. They are to be encouraged to undertake project activities related to any aspect of creation of ICT environment.

Conclusion

A long range ICT vision of Agricultural Universities has been outlined in the present paper. It is not possible to implement the vision at one go. The impediments are (*i*) lack of financial resources at one time, (*ii*) lack of manpower to spearhead the implementation, and (*iii*) lack of preparedness on the part of faculty and management to adapt to a total ICT environment. What is required is a realistic planning of implementation of ICT applications over a time frame in phases. The long range ICT goals need to be broken down into short range goals identifying the resources required so that resources are arranged in time and used efficiently. However, provision of computers at all workplaces and at least a partial networking of the campus with a reasonable bandwidth need to be implemented speedily to start with.

Implementation of ICT is not simply an introduction to new technology, but a complete rethinking on how the university's functions are achieved. Working habits of people are to be changed for success.

It is imperative to have a perpetual training system for all (including teachers, students, managers and support staff) for sustenance of the benefits accrued out of ICT implementation. Every body has his/her share of responsibility to maintain, manage and enrich ICT environment in the campus.

The aspirations of students to work in an ICT environment make them the most ICT friendly community in the campus. They should be involved in various ICT activities depending on their competence. No institute of higher learning can manage its ICT environment without the active participation of its students.

The world of learning is basking under the ICT sun. We have to catch up. It is now or never.

Chapter 6

Value Added Agricultural Education: Issues and Strategies

India has developed strong agricultural education system since independence. The establishment of State Agricultural Universities on the pattern of Land Grant Colleges/Universities of the United states contributed to reorganization and strengthening of agricultural education system in the country. The green revolution with its impressive social and economic input would have not been possible without the contribution of these centers of learning in the form of development of trained scientific manpower, the generation and assessment of new technologies and their dissemination to the farming community. The human resource developed in agriculture education system has been instrumental in agricultural transformation in the country. However, the agricultural education system has not kept pace with the rapid technological development taking place globally. During the early part of establishment of SAUs the academic programmes were structured to produce graduates who became primarily technological agents. The steps are being taken to improve the human resources in agriculture, but the glamour of government jobs dominates in our thinking. In recent years, private sector has begun in big way to absorb some of our graduates. It is opportune time for many to become entrepreneurs the present era of specializations and development in modern science, it is necessary that we restructure our agricultural education in a manner that the graduates coming out are analytical, technology oriented, sensitive to social and economic issues of farming in India, and able to meet challenges of new millennium with determination, zeal and commitment. The existing system of agricultural education is not fully equipped to enable agricultural graduates to capture other emerging job opportunities in private sector. Keeping the above in view there are some of the recommendations:

Value Added Agricultural Education

Existing agricultural education system in the country is unable to meet the futuristic need of the changing clientele. The quality of existing agricultural education system is not able to equip agricultural graduates for multifarious jobs which they are expected to take up. Thus, they do not meet the requirements of diverse employment opportunities being generated. Now even the very sustainability of existing agricultural education system in under question. While the students are lacking interest in agriculture and teachers are unable to attract the, the market for agricultural graduates does not appear to be dynamic enough to help sustainable placement of agricultural graduates. Looking to the lack of motivation of students of agriculture on one hand and significance of agriculture in the national economy on the other, agriculture degree should be made technical by changing nomenclature at national level so that due recognition is given to the degree as professional one and good students are attracted to the noble profession. The existing instructional delivery system, admission and curricula of SAUs vis-à-vis national perspective based on SWOT analysis should be reviewed. All the stake holders eager to improve the existing system of agricultural education should be involved in formulation of course curricula and methodology. ICAR should be approached to provide necessary infrastructure to develop a model for dynamic system of educational technology for improving agricultural education in the country as per present day requirement.

Quality Teaching for Enhanced Learning

Quality improvement in agricultural education is directly influenced by the aptitude and capacity of the teacher. A teacher must continue to learn. An institutional provision is required to be made where in the teachers are subjected to the process of upgrading their knowledge, practical skill and competence. Up gradation and improvement of practical skills of teachers in their respective specialized domain should also be done on priority, both within and outside the university and by inter-university collaboration. Likewise for developing skills in teaching similar facilities may be created for the teachers. Better infrastructure must be created to update the laboratory and field facilities. This is essentially required to be at par with international standards in the changed scenario under W.T.O requirements of globalization. Course curriculum must be revised to incorporate the requirements of employers at both national and international level. The requirements of the industries and other stakeholders should also be kept in view while doing the exercise. The up gradation of the entire programme must include the latest innovations and the use of modern tools such as Biotechnology and Agricultural Information Technology in accordance to have sustainable food, nutrition and environmental security. There is a need to establish a strong student's counseling and placement cell in every university which the ICAR should accordingly fund for its infrastructure and human resource needs. The better placement will certainly attract better students and generate the interest in Agricultural Education which at present is at lower ebb. Practical crop production should be further strengthened and students must be involved in alternative related sectors like poultry management, dairying, food processing, plant protection, nursery management, floriculture, farm machinery, sericulture, apiculture, mushroom culture and also bio-fertilizers etc as these are the sectors where many of them can find self-employment in contrast to crop production. The overall curriculum is required to be framed in order to make it job oriented. In order to achieve better placements for agricultural graduates, agricultural universities should establish strong students' Counseling and Placement Cell with ICAR's support for infrastructure and human resource needs. The better placements will certainly attract better students and generate the interest in Agricultural Education which at present is at lower ebb. While new avenues of jobs opportunities are to be explored the existing vacancies in universities, ICAR institutes central and the state line departments should also be filled-up on priority basis.

New Approaches for Value Added Agricultural Education

To make agricultural education effective teachers should be effective in teaching practices. Teachers should have personality traits that are associated with effective instruction and enablement of students to achieve specific tasks. Teacher needs to be facilitators not an intimidating entity. Mortern methods of information and communication technology be used. There is need to establish an excellent teacher's training institute. Faculty hired should be of premium quality and offered consumerate contractual salaries. Excellent teachers be identified as national teachers for their teaching ability. They should be allowed to travel different universities for teaching and imparting training. Facilities should be created in each universities for preparing ICT friendly teaching module. Large number of lecture halls be prepared with LCD, computer and be internet enabled. In order to bring improvement in quality of teaching, there is need to bring ablout improvement in evaluation system. At present evaluations are taking place causally due tom lack of staff and resources. Special cells can be developed to train people in designing student evaluation tools to test their power of observation, draw inference, data analysis, problem solving capacities.

Sharing Teaching Learning Experiences

Teachers are selected for their competence in subject matter, but hardly get any training on effective delivery of subject matter. There is little or no discussion in SAUs regarding teaching methodology. Teaching methodology and pedagogy have a science of their own which needs to be mastered. A lot of experimentation is required to be done to see that agriculture education is delivered effectively. It is very essential that teaching be imparted in more logical and organized manner rather than simply overloading information.

Chapter 7

Excellence in Agricultural Education: A Case of Pantnagar

Agriculture is a major source of sustenance and way of life. It sustains the livelihood of more than 650 million people in India. The rapid development in agriculture leading to food surplus from a food scarcity was possible due to the development of effective human resource in agriculture leading to the green revolution. In addition to the major source of sustenance, the agriculture sector is playing an important role in ensuring employment generation and economic growth. The effective human resource in agriculture was possible due to the broader outlook of the Education Commission (1948-49) known as Radhakrishnan Commission who realized that being the majority of the people of India live in villages, rural development would be necessary for the development of the country. Thus they made the suggestion of establishing rural universities in the country with a view to providing practical education for development of rural India. However no progress was attained for a long time until the Chief Minister of Uttar Pradesh, Pandit Govind Ballabh Pant initiated steps for the establishment of the rural university. Being impressed with the developments in agriculture in USA he deputed a team consisting of Shri. A.N. Jha Chief Secretary of UP and Harpal Singh Sandhu, a farmer settled in Tarai area of UP, who recommended the establishment of an Agricultural

University on Land Grant pattern of USA. Following this recommendation another team comprising of five Indian members, called the Indo American Joint Team to suggest setting up agricultural universities in India, was sent to USA in 1955. The team recommended setting up of rural universities in UP (Tarai region), West Bengal, Bihar, Orissa, Travancore-Cochin and Bombay State (Anand). As a result, consequently in 1960 the first and foremost agricultural university of the country was established at Pantnagar as **Govind Ballabh Pant University of Agriculture and Technology** which was dedicated to the Country by Pt. Jawahar Lal Nehru, the first Prime Minister of India.

In the mean time, Cummings Committee recommended setting up of post graduate school at IARI and IVRI. Indian Agricultural Research Institute was recognized as a Deemed University under the UGC act of 1956 and since then started awarding Masters and Doctorate degrees in the arena of agriculture and allied sciences.

The human resource developed in agriculture education system has been instrumental in agricultural transformations in the country; however it could not keep the pace with the rapid technological developments taking place globally. In fact, globalization has affected tremendous change to the character and function of education worldwide. Modern developments in agriculture has made the higher education in agriculture more demanding for inculcating recent knowledge in agriculture, skill and attitude in youths opting for the career in agriculture. The modern agricultural education should focus in such a way that the graduates passing out are analytical, equipped with recent technologies, sensitive to socio-economical issues of national and international farming community and confident to meet the global challenges of new millennium with determination, zeal and commitment. In the light of the depleting land resources, genetic biodiversity, injudicious exploitation of natural resources, ever-increasing population *vis-à-vis* demand for food and other commodities, global warming, squeezing chances of employment etc. the future strategy for agriculture particularly education should be focused to produce the quality graduates who are self confident and competitive nationally and internationally to make the agriculture sector further stronger.

G.B. Pant University of Agriculture and Technology: Role Model in Agricultural Education

G.B. Pant University of Agriculture and Technology, Harbinger of Green Revolution, enjoys the status of first agricultural university of the country truly established in the midst of the rural area and recipient of the Best Institution Award twice in 1997 and 2006. The University has also been provided with special grant of Rs. 50 crore for Institution of Excellence. The University boasts for having very strong infrastructure to cater the needs of teaching, research and extension, the three pillars of agriculture development. The infrastructure related to agricultural education encompasses very large University farm, model Crop Research Centre, Vegetable Research Centre, Horticultural Research Centre, Seed Production Centre, Mushroom Research and Training Centre, Pantnagar Centre for Plant Genetic Resources, Model Floriculture Centre, Medicinal Research and Development Centre, Agroforestry Research Centre, Instructional Poultry Farm, Instructional Dairy Farm and the vast residential campus area for students, staff members and faculty. Besides, the University has number of research stations and KVKs throughout the state covering all representative situations. In College of Agriculture Central Analytical Laboratory facility is being developed and equipped with ultramodern facilities. Besides, three Smart Class Rooms are being established in the college as a progressive up-gradation to fulfill the need of an hour. In mid 1960s and early 1970s the University was instrumental to engineer the 'Green Revolution' with the massive efforts of the scientists and faculty in association with scientists from other institutes and farmers. This led in dramatic transformation in the concept and process of agriculture in India. As such broadly speaking the administration and supervision of agricultural education in India is entrusted to the provincial State Governments and Central Government. However, overall day to day administrative functions are looked after by the Dean, Agriculture Faculty under the overall guidance of Vice-Chancellor. The Dean disposes the various functions with collaboration and cooperation of the Head of Departments, In-charge Time Table and Faculty Secretary. The College of Agriculture comprises 11 Departments and is a constant source of high quality human resource. It also serves as a valuable resource contributing to national and/state agricultural policies.

Special Features

Student Advisory System

At UG level, immediately after registration, the students in group of 5 are attached to an individual teacher who acts as an advisor to students. During the course of study of students he (advisor) remains in close contact with the individual students and advises them for various academic activities, helps in planning, guides, mends and nurtures them for all sphere development.

Course Curricula

The GBPUA&T, Pantnagar has played a role model in agricultural education by providing innovation in development of course curricula, teaching methodology, faculty development etc. The Faculty of Agriculture at Pantnagar has contributed significantly in developing the course curricula at National level through IV Deans committee. The course curricula as per the IV Deans' Committee report have been implemented with the minor modifications to meet local/state requirements. It consists of three components namely, basic supporting science and foundation subjects, general education subjects and agricultural subjects for comprehensive understanding of agriculture as socio-economic system.

Among others the Work Programme and Practical Crop Production are the special features of course curricula at GBPUA&T, Pantnagar. The Work Programme is offered in the first semester itself which gives an opportunity to all students to get an essence of doing work together in the fields, remove hesitancy of working with agricultural equipments in fields, to get the feeling of equality and develop team spirit etc.

The college has been the pioneer in starting the Practical Crop Production (PCP) course in 1962 for its first batch. The PCP course is unique in it and provides an opportunity to students:

- ☆ To gain an experience in planning, execution and cost-benefit analysis of diversified and modern techniques in the field of crops, mushroom, horticulture, floriculture, vegetables, food processing and apiculture enterprises/ production

- ☆ To work in Plant Health Clinic for diagnosis and forming recommendation for various problems in plant
- ☆ Inculcates team spirit and dignity of labour
- ☆ Experiential learning through earn while you learn
- ☆ Unique model for practical training in agriculture and has since been adopted in several agricultural universities in the country.

An opportunity, for application of the knowledge acquired during academic theory and laboratory classes and building the confidence, is offered to the students through Rural Agricultural Work Experience (RAWE) during the first half of the fourth year of the programme wherein the students are assigned to work in the real fields *i.e.* framers' fields, research stations, industries under the close supervision of the expert teacher/faculty. This approach is essential for preparing the students more competitive, confident for their future career. Thus RAWE course programme is focused on project formulation, industrial attachment and plant health clinic.

In this competitive and demanding world, particularly in private sector, hands on training in the area of interest of the student are must. To fulfill this requirement a package of courses as 'Experiential learning' is offered during the post half of the IV year. The package of experiential learning comprises varying courses under different modules *viz.,* crop production, post harvest technology and value addition and agri-business management. Out of the various modules, students have the freedom to choose any two modules of his interest for total 20 credits. PCP and RAWE in addition to National Social Service (NSS) aim to develop the leadership quality among students and personality development focusing on commitment, knowledge, skill, aptitude, and attitude.

Course Planning

The effective delivery of instructions during the course of study is pivotal. After the registration of the students for the programme, usually the whole batch of one academic year is divided in to three sections *viz.,* A, B and C etc. with almost equal number of students. Each section is allotted a set of different instructors for delivering the instructions in various courses. For uniformity and equal coverage of contents the course planning type of workshop takes place among the instructors of different sections. Group discussions are organized during the course with the assistance of the panel of selected students who initiate this activity under the guidance of Course in-Charge Instructor. For ensuring the quality delivery of the instructions and overall development of the students senior faculty members and/or Dean Agriculture himself monitors various activities by visiting the on-going classes from time to time. The Under Graduate as well as Post Graduate academic programmes are very impressive and effective as there are 100-105 instructional days in a semester and 85 per cent attendance is compulsory for appearing in examination.

Information and Communication Technology

Information and communication technology is gaining importance as a backbone of acquiring knowledge, skill and communication. It has played an instrumental role in modernizing delivery of agricultural education. Information revolution has provided solutions to several challenges of quality teaching-learning process. Use of computer has revolutionized greatly all aspects of information browsing, handling and delivery; modeling, simulation of agricultural practices and epidemics of plant diseases, generation of data bases, development and use of expert programmes, crop cultivation practices, diagnosis and decision related to agricultural activities. At GBPUA&T, to cater the need of making the students sufficiently competent and able in ICT a Central Computer Facility (CCF) has been established where students are trained and allowed to surf the information and communication (e-mail) through the individual user ID.

Extra-curricular Activities

Students are facilitated and encouraged to participate in various extra as well as co-curricular activities *viz.* literary, cultural and sports. The University has a central stadium consisting gymnasium, indoor as well as outdoor game facility; and the coaches to guide. Besides, every hostel is equipped with indoor game and entertainment facilities. Agricultural Society provides a platform to every student(s) for nurturing the professional skills in the field of individual student's interest through debates, quizzes, exhibitions etc. The society also runs a wall magazine to provide a platform for training as well as to fulfill the quench for writing literary and technical articles. '*Oasis*' a monthly newsletter highlighting the campus activities is solely articulated and brought out by the students.

Placement Bureau

The Placement Bureau of the University facilitates the students for employment through building the purposeful linkages between the students and the prospective employers. Number of employers performs the campus interviews and absorbs the graduates and post graduates. Every year of students get suitable employment through the Placement Bureau. Alumni of the college are holding the key positions in administration, management, consultancy, industry, education, research, banking and other fields in the country and abroad. To meet the specific placement needs the college has its own Placement Cell. It also organizes the regular communication and personality development trainings to equip the students for the job market.

Future Outlook

Agricultural education is empowerment of knowledge, skills and confidence to manage it. The revolution in Information and Communication Technologies (ICTs) has transformed the outlook of the stakeholders towards agriculture. The farmers are tempted towards commercial agriculture and steeping towards professionalism in agriculture in place of simple hand to mouth agriculture. In the light of this the trained manpower coming out of the colleges/universities should be competent enough to take care of it and deliver effectively. To achieve this, although academic programmes in agriculture have been recently revisited and modified by the IV Dean's Committee, there is an utter need to follow it and implement particularly RAWE and Experiential Learning, on the ground in true sense.

The development and availability of technologies playing a vital role in transforming steadily an instruction-centric university model to leaner-centric integrated network model based on access to learning resources and students' initiative. To facilitate this it is imperative to develop basic ICT infrastructure to realize the advantages of the application of the technology step by step to bring more and more of academic activities under the ambit of ICT. To say, development of Smart Classroom having a provision of internet connectivity makes the whole world shrink in classroom. Every link in the system of higher education in agriculture should be trained on a continuing basis to equip them with working skills to exploit the ICT in their and system's benefit optimally.

The human resource in agriculture in recent years is expected to have professional approach. To build up this the value addition in terms of practical skills, group discussion, panel discussion, analytical ability, confidence and ability to handle varying situations and exposure and expertise in ICTs are most vital and thus demands serious attention.

Section 2
Educational Technology

Chapter 8

Educational Technology: Concept

Everyone is talking about educational technology in the context of educational reform in present era. The challenges of education call for technological intervention to improve quality of learning. It is the technological accepts of instruments or media to be used in education or a branch of education to deal with social, psychological and pedagogical aspects of teaching learning. Simply put the queer combination of the terms 'education' and 'technology' create more confusion. Probably the discussion about the term became more professional with availability of various audio-visual aids –earlier slides, overheads, flip charts, graph pictures, audio tape, film, projectors and later radio, television, computers, multi-media, internet to integrated technologies together called as instruction media, they require planning, design, production and testing to make instruction effective. That is why the term educational technology refers to various considerations about learners. Learning resources, learning abilities, levels, situation, objectives etc. both technical and human aspects are to be considered to provide a learning solution. U. S. commission on instructional technology started "Education Technology" is a systemic way of designing, implementing and evaluating the process of learning and teaching in terms of learning and communication and employing a combination of human and non-human resources.

Education technology focuses on learning as an outcome. In order to ensure learning various factors of teaching and learning are considered in the light of researches different human and technical disciplines. in the process, suitable technology are used. Thus, educational improvements are based on rational and imperial findings rather than trail and error.

Thus, instructional media like TV, Computer, Video or Internet are only means to the end. They help in motivating learners and presenting materials systematically with the help of audio-visual stimulate. They can be used again and again. A teacher can conveniently facilitate discussion and help in learning effectively United States National Council educational Technology described four characteristic of educational technology.

- ✰ Clear definition of objectives to be achieved by learners right in the beginning the expected learning outcome should be spelt out in certain observable terms so that appropriate decisions can be initiated to achieve.
- ✰ Application of principles of learning to analysis and structuring of subject matter to be learnt learning does not happen randomly. People learn differently and content has been identified structured and presented keeping the principles of learners in visual to produce learning out come.
- ✰ Selection and use of appropriate media for presenting messages/materials.
- ✰ Media help in structuring content, providing various stimuli and explaining the content in various different ways. Choice of media depends on content, objectives, learners, etc. thus, a rational and pragmatic choice has to be made in a give situation.
- ✰ Use of appropriate methods for evalution of materials must be planned before hand to make sure that learning is tested through valid tools.

All these make the task of teaching scientific, imperial and effective. A teacher becomes experiments in his own right.

Educational Technology movement started in the country with setting up Audio-Visual units for preparation and use of Audio-

Visual aids in the class. Later Education Technology was launched and centre for Educational Technology was setup in NCERT. Later education Television programmers were started. Research on teaching, learning, media use, distance learning and models of teaching were undertaken by Educational Technology Cells in Universities. In agricultural education use of Audio-Visual aids and other media well encouraged after setting up of SAUs. However, teachers' training in use of instructional media was started after setting up a National Academy of Agricultural Research Management.

Chapter 9

Educational Technology in State Agricultural Universities: Past, Present and Future

The country has developed strong agricultural education system over a period of nearly four decades and a half. The establishment of State Agricultural Universities on the pattern of Land Grant Colleges/Universities of the United States contributed to reorganization and strengthening of agricultural education system in country. This has certainly helped in ushering the country from the era of food scarcity to food surplus. A rapid stride made in Agriculture, during last part of the century was significantly contributed by the development of effective human resource in agriculture. Scenario of agriculture has changed, dimensions of problems are different, than what was before a decade or two. The human resources developed through agricultural education system have been instrumental in agricultural transformation in the country.. A large network of agricultural education system consisting of 38 State Agricultural Universities, 5 Deemed-to-be Universities, 3 central Universities with faculties of agriculture and 1 Central agricultural University has contributed substantially to give a scientific base to Indian agriculture. The challenges facing Indian agricultural education are different than before. With the liberalization of economy, a number of multi-national companies with high tech agriculture are entering into agriculture and the competitiveness is increasing. The question before us is how to improve our agricultural education and make our graduates competitive nationally and internationally. There has been widespread consensus that quality of graduates in various disciplines of agricultural sciences does not match with the changing needs. With emerging job market scenario in front, introduction of subjects and methods of teaching and learning aligned fully with employability should form the core of the strategy for agricultural education. While doing that needs of self-employment seekers, perspective employers and changing scene in agricultural trade and environment management need to be intertwined with the model competencies and skills. Employment opportunities in the off-farm sector are expected to increase at a faster rate than in agriculture. This will further emphasize the present employment shift of agricultural graduates to related sectors, requiring a revision of existing curricula to better address educational needs. Furthermore, agricultural education curricula need to be redirected to more specifically address national problems. This reorientation should incorporate both the new role of market-oriented agriculture as well as issues of direct relevance to the improvement of subsistence agriculture and rural poverty. Indian Council of Agricultural research has introduced many changes to revamp education system in general by introducing various components such as educational technology human

Resources development, modifications of under-graduate and post-graduate curriculum, uniform pattern of curriculum and grading, computer facilities, upgraded laboratories, etc. In the era of WTO, it is necessary to improve the quality of teaching for professional development of teachers as well as students. It is here that science of educational technologies comes in picture to analyze the process of education and solve the problems of teaching and learning this call for experimentation with methods of teaching for enhanced quality and professionalism. Advances in educational technologies must now be used to systematically design, implement and evaluate educational process. Thus, it is necessary that improvement programmes are run to improve quality of teaching sooner than later. Thus the following components need our immediate attention:

Human Resource Forecasting

The rising unemployment in agriculture science is sure sign of lack of proper planning and development. The rising number of state agricultural universities and private colleges indicate mismatch between demand and supply. While there is definite need to restrict admissions to under-graduate and post-graduate courses to a limit, the sub-standard private colleges must be closed. Assessment of human resource requirement in agriculture and forecasting of demands in particular fields is another need of the time. To change over to an effective market driven academic system, it is essential to predict changes in the job market. This calls for active involvement of user groups in planning and management of educational programmes.

Designing and Development of Courses

With emerging new job market in front, it is now felt that innovative courses and methods of teaching need to be formulated sooner than later. Practical job-oriented, competency based courses in organic farming, intellectual Property Right, national trade and export, natural resource management, rural communication, NGO management, Post-harvest technology, bio-diversity conservation, agricultural heritage, Designing competency based courses needs training in the craft which is scarcely available in the countries. Apart from the content, teaching methodology and evaluation procedures need to be specified to achieve the goal of competency based courses areas. This is an area calling for advanced training.

Use of Information and Communication Technologies

With advances in information and communication technologies, most traditional methods and media of teaching have become obsolete.Infrastucture, knowledge and Skills in this highly specialized area is not available in the country. Scattered efforts are being made to initiate distance education, on-line learning and use of internet for agricultural education, but not much infrastructure and expertise is available within the Nation Agricultural Research system. Thus this calls for special attention

Action Learing

There is an ongoing revolution in the field of teaching and

learning especially in the professional fields to the extent that classroom teaching is becoming obsolete in many professional universities of the world. This calls for innovations. No doubt, classroom teaching has served its purpose in the past but preparing graduates to diagnose, solve problems and adjust with different types of work conditions calls for use of innovative methods hitherto used in others, action learning alone can expose them to the intricacies of interrelated problems in the field. Hand on training is needed for there should be continuous opportunity to apply and test skills.

Centre for Instructional Resources

Each SAU needs centre for in In-house training of teachers in teaching skills. Beginning with new teachers, the emphasis would be to impart skills of effective teaching and disseminate best teaching practices to all the teachers. Besides, support would be provided to teachers in preparing teaching packages in digital form for standardizing teaching practices, providing help and guidance in preparation of portfolio of interested teachers, formation of quality circle and conducting action research. Facility for video recording of teaching to interested teachers and preparations of literature for quality teaching can be made available there.

Partnership with Other Institutions

Not all the SAUs are endowed with vast infrastructure to provide practical education of relevance to the job market. This means that liaison needs to be created with the government and private organizations to forge working relationship for meaningful association. Electronic net working between SAUs ICAR institutes, National Research Centers, Private and Non-Governmental Organization is called for to make agricultural education useful. Regular interaction with the private and industrial houses as well as the government organizations can help academics keep in touch with the changes and include them in the teaching process. Frequent interface with the employers and other stakeholders can sensitize both the parties and make education relevant. This calls for organizational linkages and networking.. The necessity for better and closer links between education and employment in order to meet the actual needs of the region or the country through an adaptation of curricula and teaching methods to meet these needs.

Nature of Collaboration

The nature of collaboration required with Land Grant Universities in USA to provide training, infrastructure support and exchange programmes to introduce different aspects of educational technology in SAUs:

- ☆ Training of teachers in educational Technology
- ☆ Development of Infrastructure
- ☆ Exchange of students
- ☆ Teaching & Research programmes in educational technology

Chapter 10

Education Technology Cell: Concept and Future Outlook

Educational technology is the buzz word today in the context of qualitative improvement in teaching and learning. The challenges of education call for technological interventions to improve quality of learning. Several efforts were made in the past few years to enhance quality of teaching and learning through curriculum up gradation, inclusion of new courses, students' evaluation of teaching, organization of several workshops for reforms in education systems, introduction of external examination and close monitoring. In order to consolidate the gains on continuous basis and learn from the theories of education, each SAU needs to establish an educational technology cell to improve teaching through training and dissemination of best teaching practices, preparation of teaching packages in digital form for standardizing teaching practices, help and guidance in preparation of audio-visual media,, formation of quality circle and conducting action research, etc. This Cell may carry out activities as below:

- ☆ Evaluation and researches for up-gradation of curriculum, instructional technologies, teaching methodology, psychology of teachers and students etc.
- ☆ Serve as knowledge repository and clearing house for disseminating information on different aspects of teaching and learning

- Regular publication of journal, resource materials, tips for teaching and reports as ready-reckoners for teachers for enhancement of effectiveness of teaching.
- A documentation centre for books and literatures on educational technology
- Training and workshops of students for enhancement of their soft skills, analytical ability, learning ability etc.
- Action researches by teachers on teaching &learning
- Innovations in teaching methodology
- Consultancy to younger teachers on problems in teaching and learning.
- A data bank for evaluation results, feedbacks and inferences over curriculum, teaching methodology and teachers.

Humble beginning has been made at Pantnagar with availability of expertise and infrastructure but there is need to think about organizational structure, budget, training, hardware and other related aspects in view of the emerging developments.

Initiatives at Pantnagar

Looking to the needs for qualitative improvement in teaching, College of Agriculture initiated several efforts during last one year and a half. To begin with, curriculum was upgraded and new courses were added to make curriculum contemporary. Next, consultations and interactions with teachers and students were conducted. Two brainstorming session were organized on improvement of teaching skills at college level in which, teachers and students took interest to share their concerns. Several interactive workshops were organized for the undergraduate students about need for qualitative changes in teaching soft skills Hon'ble Vice-Chancellor also observed and provided very pertinent remarks over the process. Later scheduling of internal examination and close monitoring of teaching was undertaken. An attempt was also made to explore parameters for evaluation of teachers. In keeping with the continued effort to improve teaching, a National conference on Value added agricultural education –Issues and strategies were conducted on Oct. 21-22, 2005. Hon'ble Vice-Chancellor has been kind enough to approve starting of education technology Cell. in the College of Agriculture with Dr.

Birendra Kumar, Professor Agril. Communication as Coordinator and Dr. S. K. Kashyap, Associate Professor, Agril. Communication as Associate Coordinator. This Cell has a vision of fulfilling long-cherished goals of educational improvement through continuous research, HRD efforts and consultancy in teaching. Now that educational technology Cell has been initiated after, the following activities are proposed.

- To organize periodical training for teachers on instructional development and professionalization of teaching
- To conduct evaluation and researches for up-gradation of curriculum, instructional technologies, teaching methodology, psychology of teachers and students etc.
- Serve as knowledge repository and clearing house for disseminating information on different aspects of teaching and learning
- Regular publication of journal, resource materials, tips for teaching and reports as ready-reckons for teachers for enhancement of effectiveness of teaching.
- Training and workshops of students for enhancement of their soft skills, analytical ability, learning ability etc.

Plan for Future

It is planned to create a teachers, room in the college in form of "Educational Technology Cell" where the ambience may inculcate the values and instinct of improvement in dedication towards teaching. The plan is to give:

- A motivating model class room
- A documentation centre for books and literatures on educational technology.
- A clinic for personal consultancy on specific problems in safe and protected atmosphere.
- A data bank for evaluation results, feedbacks and inferences over curriculum, teaching methodology and teachers.

Chapter 11

Experiential Learning in Teaching

One can learn from experience,
but one cannot be taught by it.

- Adam Phillips, *On Kissing, Tickling and Being Bored*

The great philosopher **Socrates** believed that role of teacher was that of "drawing out of the student" and bringing to the level of conscious recognition that which he already knew. Socrates developed a method of teaching by questioning. He invited student to teach him, to give him the best answers to the questions he posed; and then helped them to examine critically their own answers to modify them and finally developed adequate understanding to develop truth. Socratic method of teaching is generally recognized as an educationally sound procedure. Many philosophers from Aristotle to John Dewey recognized that we "learn by doing".

The medieval belief about teaching was that we could teach people by drilling holes in the human head and, with a funnel, pour information into the brain. In most of our educational institutions, for teaching the model of passive learning is used and students are passive participants in the teaching- learning- process. In today's competitive world the teaching is a challenging task, students should no longer be the passive recipients of knowledge but integrating teaching which stimulate senses of students for active learning is the need of the hour. Active learning results in longer-term recall, synthesis, and problem-solving skills than learning by hearing, reading, or watching. Education needs to move from a learning-by-telling model and even learning-by-observing (**as in the case-method**) to a learning-by-doing model. There is need to move from passivity to activity and teachers must learn to extrapolate from their experiences and see how to apply what they've done to new instances.

Students can learn more readily about things that are tangible and directly accessible to their senses-visual auditory, tactile and kinesthetic. With experience they grow in their ability to understand abstract concepts, manipulate symbols reason logically and generalize. These skills they develop slowly when concrete examples about new ideas are given. Concrete experiences are most effective in learning when they occur in the context of some relevant situations.

Experiential learning in education is advocated by many of the contemporary education thinkers which emphasized on "Learning from experience" which might be called "nature's way of learning". It is "education that occurs as a direct participation in the events of life" **(Houle, 1980, p. 221, quoted in Smith 2003).** It includes learning that comes about through reflection on everyday experiences. Experiential learning by oneself is also known as"informal education" and includes learning that is organized by learners themselves.

Experiential learning has come to mean two different types of learning:

1. *learning by yourself* and
2. *Experiential education* [experiential learning through programs structured by others] **(Smith, 2003).**

Experiential Education

Experiential education is learning through programs and activities structured by others. In experiential education teacher's role is to design prescribed learning exercises based on student' experiences which are based to transmit knowledge and information for pre decided goal. Principles of experiential learning are used to design of *experiential education* programs. Emphasis is placed on the nature of participants' subjective experiences.

Some Viewpoints about Experiential Learning

Kurt Lewin (1951) stated that learning involves feeling things about the concepts (emotions) and doing something (action). Lewin who stressed the importance of a people being active in learning. Jean Piaget described intelligence as the result of the interaction of the person and the environment.

Cranton(1994) call it "learning by doing"; "action learning"; "action science"; "reflection in action"; and "praxis" etc.

Praxis is the Greek word that means action with reflection. (Praxis = Experience + Reflection > Action.) In educational situations, we describe, analyze, apply, and then implement our new learning. When we practice a skill, analyze our practice, and then repeat the practice at a higher level, we move practice to praxis. *We learn what we're doing.*

Kolb,David. (1984) describes learning as a four-step process. He identifies the steps as (1) watching and (2) thinking (mind), (3) feeling (emotion), and (4) doing (muscle). He draws primarily on the works of Dewey (who emphasized the need for learning to be grounded in experience), Kolb wrote that learners have immediate concrete experiences that allow us to reflect on new experience from different perspectives. From these reflective observations, we engage in abstract conceptualization, creating generalizations or principles that integrate our observations into sound theories. Finally, we use these generalizations or theories as guides to further action. Active experimentation allows us to test what we learn in new and more complex situations.

Rogers, Carl. (1969) experiential learning is equivalent to personal change and growth. Rogers feels that all human beings have a natural propensity to learn; the role of the teacher is to facilitate such learning. This includes: (1) setting a positive climate for learning, (2) clarifying the purposes of the learner(s), (3) organizing and making available learning resources, (4) balancing intellectual and emotional components of learning, and (5) sharing feelings and thoughts with learners but not dominating. According to Rogers, learning is facilitated when: (1) the student participates completely in the learning process and has control over its nature and direction, (2) it is primarily based upon direct confrontation with practical, social, personal or research problems, and (3) self-evaluation is the principal method of assessing progress or success. Rogers also emphasizes the importance of learning to learn and an openness to change.

Roger's theory of learning evolved as part of the humanistic education movement ***(e.g.*, Patterson, 1973; Valett, 1977).** Roger's theory of learning originates from his views about psychotherapy and humanistic approach to psychology. It applies primarily to adult learners and has influenced other theories of adult learning such as Knowles and Cross. Combs (1982) examine the significance of Roger's work to education. Rogers and Frieberg (1994) discuss applications of the experiential learning framework to the classroom.

Borzak (1981) noted experiential learning involves a "direct encounter with the phenomena being studied rather than simply thinking about it." That is, students were facilitated to 'learn by experiencing' rather than by simply sitting and listening.

Kolb(1984) provides one of the most useful descriptive models of the adult learning process available, inspired by the work of Kurt Lewin. This suggests that there are four stages which follow from each other: **Concrete Experience** is followed by **Reflection** on that experience on a personal basis. This may then be followed by the derivation of general rules describing the experience, or the application of known theories to it **(Abstract Conceptualisation)**, and hence to the construction of ways of modifying the next occurrence of the experience (**Active Experimentation**), leading in turn to the next **Concrete Experience**. All this may happen in a flash, or over days, weeks or months, depending on the topic, and there may be a "wheels within wheels" process at the same time.

The conclusions drawn from these theories are that teacher should plan learning experiences so that students can:

- ☆ Perceive the information;
- ☆ Reflect on it, how it is going to effect their life;
- ☆ Compare it to how it fits into their own experiences;
- ☆ Think about how this information offers new ways for them to act.

The curriculum for agriculture education is varied in nature. It includes disciplines from the pure sciences like physics, chemistry, and mathematics to applied behavioral sciences like sociology, psychology and extension education with agriculture subjects. Thus, teachers in agriculture education are required to integrate principles of experiential learning so that agriculture students are able to apply what they have learned as being useful beyond the classroom. There is need to integrate practical examples in the teaching of agriculture and it may be aimed to problem solving in real life situations.

Why Do we Need Experiential Learning

Jerome Bruner noted in 1966 that "the purpose of education is to stimulate inquiry and skill in the process of knowledge getting, not to memorize a body of knowledge". Kolb places learning into three categories:

- ☆ Experiential–"a holistic integrative perspective on learning that combines experience, perception, cognition and behaviour".
- ☆ Rationalist and Other Cognitive Learning–primary emphasis to acquisition, manipulation, recall.
- ☆ Behavioral–denies any role for consciousness and subjective experience in the learning process–*e.g.* learning a poem by rote, learning how to overhaul a carburetor.

Steps for Planning Experiential Learning

Know your Learners

The first step in planning any learning experience is that as a teacher one must be aware of the mental level, past experiences, background and skills which learners have. Depending upon the type of learners whether farmers, rural youth, farm women or university students the learning experience has to be planned. The learners may be classified on the basis of their learning styles as activist, reflector, theorist and pragmatist **(Honey and Mumford 1982)**

Learning Environment

It is important to know for what type of educational environment the learning activity is planned. Is it in the classroom, in field situation, in the village or outdoor activity? The teacher may explore the infrastructure and available resources for that particular learning environment.

Specify Learning Objectives

The experiential learning experience should relate to the learning objectives which are focussed on broader goals of the curriculum. The teacher may define what type of change is desired whether it will bring change in their knowledge, skill or some behavioral change according to it the teacher has to plan the activities.

Structure Learning Activity

The learning experience may be so structured that they are close to real life experience or where students are directly involved or relating theory which is focussed on reflective thinking where students can relate it with relevant concepts. They develop deep insight about the topics covered and are able to synthesize their thoughts and apply them in real life situations.

Define your Own Role as Facilitator

Carefully record the student's reflections on particular subject. Ask the students to briefly write their experiences. What important points emerged from the whole process? How do they feel about it? Get their feedback. Try to build up the connections with their earlier experiences.

Reconsideration

On the basis of the students feedback and analysis of the whole process the implemented experiential strategy can be modified and new exercise and experience can be further designed and linked with new concepts. Thus this process can be continuous.

Some potential benefits are:

- ☆ Broadens and strengthens learning outcomes.
- ☆ Builds opportunities to link with the real life experiences and enhances problem solving abilities.
- ☆ Promotes learning as a lifelong process.
- ☆ Integrates theory and practice which fosters the ability of students to grasp more deeply fundamental concepts and their importance.
- ☆ Allows students to "capitalize on practical strengths while testing application of ideas discussed in the classroom".

- Focussed on relevance and application of ideas which are demonstrated and tested against their accumulated experience and wisdom.
- By relating theory to practice, produces learning in a more familiar environment and therefore fosters more productive learning.

The experiential learning experiences will be useful for teaching life skills, sensitizing towards some social issues, developing team projects, linking students with organizations, reemphasizes experience as a critical part of learning, permit methods to follow up the experiences so that students can apply the insights gained through additional relevant experiences and it help them to analyze practical problem.

Steps in Experiential Learning Sycles

Experiential learning is about learning from experience. The characteristics of the approach are:

- Learner- centeredness
- Allows the individual participants to manage and share responsibility for learning with their teachers.

The McCaffery (1986) model of four step experiential learning cycle (ELC) is an effective training/teaching strategies, which incorporate experiential learning approaches provide opportunities for a person to engage in an activity, review the activity critically, draw some useful insight from the analysis and apply the result in practical situation.

It can be graphically be represented as follows:

The First Step

The **Experience** is the initial activity of the ELC that generates information or in other words produces data. The phase is so constructed that enables participants to become actively involved in "doing" something. This may include a range of activities like the follows:

- Case studies
- Role plays
- Simulations

- ☆ Games
- ☆ Lecturettes or lectures
- ☆ Film or slide shows
- ☆ Skill practice etc.

There could be other range of techniques that can be used. It can be passive and artificial (lectures) to more active and real (skill practice). The choice of the technique largely depends upon the goal of the training/teaching session. With the completion of the activity, the next stage of the cycle is initiated *i.e.*, the **process**.

The Second Step

During the phase of the **process**, the participants

- ☆ Reflect on the activity undertaken during the experience phase, and
- ☆ Share their reactions in a structured way with other members of the group.
- ☆ May speak individually or as a full training group.
- ☆ Discuss both in terms cognitive and affective reactions that reflect the intellect and the attitude of the participants towards the activity they are engaged in.
- ☆ May try to link these thoughts and feelings together in order to derive some meaning from the experience.

Role of Trainer/Teacher

The role of the trainer as facilitator is important during each phase. During the process, he/she should be prepared to help the participants

1. Think critically about the experience
2. Help them verbalize their feeling and perceptions, and
3. Draw attention to any recurrent themes or patterns which appear in the participants' reaction to the experience.
4. To conceptualize their reflections on the experience so that they can move towards drawing conclusions.

The Third Step

The **generalization** is the next step in experiential learning cycle. In this the participants

- ☆ Form conclusions and generalizations derived from the first two steps of the cycle.
- ☆ Are helped to think critically in order to draw conclusions that might generally or theoretically apply in real life.

Questions Like

- ☆ What did you learn from all this?
- ☆ What more general meaning does it have for you?

The Instructor/trainer structures this step so that the participants work alone first, and then are guided into sharing conclusions with each other. This enables the participants to exchange thoughts and ideas so that they ay act as catalysts to one another.

Role of the Trainer/Teacher

The trainer or the teacher helps to facilitate this step by:

1. Asking and helping individuals to summarize what they have learned into concise statements or generalizations
2. "pushing back" at people to help them make their thinking more rigorous.
3. Relating the conclusions reached and integrating them into a theoretical model.
4. Making sure, within reasonable time boundaries, that everyone who wishes to share a significant insight is given a chance to contribute
5. Helping the group compare and contrast different conclusions, identifying patterns where they exist, and identifying legitimate areas of disagreement

The Fourth Step

After participants have formed some generalizations, they are guided into the **application** step of the cycle. Drawing upon insights and conclusions reached during the previous phase (and other phases), they (participants) can:

- ☆ Begin to incorporate what they have learned into their lives by developing plans for more effective behaviour in the future.

- In the ideal educational or training event, participants would be able to apply what they have learned immediately after the workshop/training/teaching ends.
- The application that they plan may relate to their profession or personal life, depending on the background and need of the specific groups.

The following technique can be used to facilitate the application stage:

- Individual work to develop a thoughtful action plan which puts thought into action
- Participants review each other plans are assist in formulating ideas for action
- Part of individual action plans are shared with the whole group in order to create a sense of joint efforts
- Additional learning needs are identified by participants
- Enable participants to be as specific as possible in developing their action plans.

It is important to stress on two other points about the experiential learning model.

1. The exact nature of each phase of the model is determined by the goals of the training sessions or the program. Once the goals are determined or defined, the sessions can be designed using the model as the framework.
2. Theory can come in two different places – either before the experience, in which case the experience becomes a way to test the theory or try out the skill implied by it, or after, when it is interwoven into the generalization phase as participants develop their own "theory". So when these two components are added the graphical representation can be presented as follows.

In Order for this Model to be Effective

- It must be rigorously applied, both in the design and delivery stages.
- Don't misuse the phrase in practice where it seems to mean letting people participate in presentation, having

questions and answer sessions after a lecture, or a role play by itself without the subsequent steps in the model.

- ✰ Most frequently, generalization and application stages are simply left out of the design or the program. As a result, the power of experiential learning is significantly diminished or is negated altogether.

Advantage of the Approach

The experiential learning model or cycle helps people

- ✰ Especially for skill training as most techniques are active, though they can also be used in some passive or lecture experiences.
- ✰ To assume responsibility for their own learning because it asks them to reflect on their experiences, draw conclusions and identify applications.
- ✰ The learner are therefore, not spoon-fed
- ✰ Nor thev are led to be dependent on experts.

Desirable

Efforts by the trainer/teacher in instructional style for its implementation effectively.

Chapter 12

Agricultural Education through Distance Learning

Introduction

Agricultural education is the key to generation and adoption of improved technologies as well as execution of policies for making strides in agricultural development. In India, agricultural higher education got a boost with the establishment of the first State Agricultural University at Pantnagar. The SAUs and ICAR institutes have made major contributions towards advancements in agriculture through generation of new types of production technologies. Self-sufficiency and export of agriculture produce have been possible in the country due to these efforts. However, there is challenge to further upgrade scientific resources in the country in specialized areas for generation of appropriate technologies in the globally competitive market.

India has developed strong agricultural education system since independence. Human resource developed in agriculture education system has been instrumental in agricultural transformation in the country. However, the agricultural education system has not kept pace with the rapid technological development taking place globally. In the present era of specialization and development in modern science, it is necessary that we restructure our agricultural education.

Hence there is urgent need for upgrading the quality of teaching in agriculture. Educational systems are dynamic in nature, continuously adapting to meet the emerging needs of society. In the era of WTO it is very much necessary to improve the quality of teaching for professional development of teachers as well as students. Education has acquired a new dimension with phenomenal breakthrough in ICTs (Information and Communication Technologies). The emergence of virtual classroom, open learning systems and distance learning has given a new perspective to teaching and learning process.

Concept of Distance Learning

Distance learning, with particular reference to higher education, is being conducted in three types of universities:

- *Conventional Universities* which deliver full time education through on-campus face-to -face delivery and draws students with homogenous features. They constitute the bulk of the educational system.
- *Distance and Open Universities* which deliver higher education via distance mode of education. Their students are not only heterogeneous in age and educational background but also employed and often unable to undertake full-time study for a variety of reasons. The students cannot devote all their time and are eager to gain degree for a variety of reasons. Technology is playing a great role in delivery of instruction in such universities.
- *Dual mode Universities* which combine both distance education and face-to-face regular mode of education. They may offer correspondence courses as well as part time degree programme.

Distance learning is an organized process that separates the student in time or place, and includes courses offered by computer transmission, television, electronic conferencing, video cassettes, discs or correspondence. It also includes face-to-face communication along with other forms of delivery. Central to the meaning of distance learning is the fact that teachers and students are separated from each other during the process of learning. Thus, effective learning in distance mode depends on a variety of media for enhancing two way communication between instructors and distant learners. There is, of course, need for fixed instructional base to administer a very systematized as well as mediated curriculum. The form of education has gone through many phases of evolution from correspondence courses off campus, part time study to 100 per cent distance education university studies.

Distance and open universities do not require class rooms and presence of students. Thus, much of the problems of presence of students and requirement for space as well as infrastructure are not there. However, there is need for systematizing educational process to the extent that variety of educational needs are met effectively with the support of media and allied services without direct interaction with teachers. This puts very heavy emphasis on scientific organization and management of distance learning system.

Many educators ask if distant learners can learn as much as students receiving traditional face-to-face instruction. Researches comparing distance education to traditional face-to-face instruction indicates that teaching and studying at a distance can be as effective as traditional instruction, when the method and technologies used are appropriate to the instructional tasks (Moore and Thompson, 1990; Verduin and Clark, 1991).Looking for possible options ICTs (Information and Communication Technologies), correspondence courses, online learning, distance learning suggest that distance learning seems to be the most economic and efficient. Hodgson (1993) notes that the key feature of open learning is that it seek to remove all barriers to learning, most notably barriers of time and place of study. The International Commission on Education for the 21st Century called for the use of distance education techniques especially in in-service training (UNESCO, 1996).

Distance Learning in Agricultural Higher Education

In case of Indian agricultural higher education, co-incidentally there is no distance/open universities solely committed to it. The programmes have to be added on the existing organizational base of regular educational system. State Agricultural Universities have been performing the triple functions of teaching, research and extension. There are separate directorate of extension education with responsibilities of non-formal education of farmers, youth and others. Correspondence courses, media broadcasts and use of print materials are quite common experiences of SAUs in extension work. G.B. Pant

University of Agriculture and technology in particular is privileged to have a department of Agricultural Communication, Directorate of Extension and Communication Centre. In fact, Communication Centre was established here way back in 1969-70 with the help and support from Ford Foundation to develop communication support for education and extension work. On the basis of the ground work done in communication, a Centre for Advanced Studies in Agricultural communication was sanctioned to develop academic programme in communication to provide trained communicators for the Communication Centre of the agricultural universities, ICAR and other research institutes engaged in similar pursuits. As a result, an infrastructure for production of audio, video and print materials has already been established. The University has been regularly sending agricultural programme to different centres of All India Radio including one in the immediate Neighborhood. G.B. Pant University already has some experiences in running correspondence courses for about two decades on modern farming for young men, women and retired personnel interested in taking up agriculture as enterprise. The programme has successfully been conducted in range of subject matter areas. The University has been pioneer in starting 'Farm School on Air' programme in collaboration with All India Radio Station at Rampur. The University has renewed the effort, of late, to provide such course in specific fields. Besides, attempts have been made to provide video news and instructional programmes on regular basis to the national network. It is expected that other SAUs are also working on similar lines. Such experiences provide a conducive background to initiate much needed distance learning programme in agricultural higher education. Every SAU has a network of Krishi Vigyan Kendras, Krishi Gyan Kendras and regional research centres to serve as counseling or support units. Keeping in view the need for strengthening qualitative instruction and training for a large population on one hand and the existing facilities as well as the experiences on the other, SAUs are fit cases for starting distance learning programme as regular feature in the most important sector of the nation. Such efforts will positively affect the quality of regular programmes of SAUs a well.

Training Needs of Agricultural Scientists

In service training of National Agricultural Research System (NARS) researchers and agricultural research managers is one way of providing and improving the knowledge and skills they require to effectively target research to the rural poor. Traditionally, in-service training relies on workshops that usually take NARS staff away from their jobs and incur high costs for facilities, travel, and subsistence of trainers and participants. In addition, workshops and structured training programmes compress training interactions into periods that are often too short to adequately impart the knowledge and skills required to improve researchers' performance. The workshops tend to be more academic and formal in nature, thus creating a large gap between theory and practice. Moreover, they often do not tap the experience potential of the scientists, researchers and managers in the rural based institutes, often distant from the centre of learning and know-how, thus reducing the two-way flow of information. They run a real risk of losing touch with the fast changing technology, which causes their skills and abilities to become obsolete and gradually deteriorate. The existing man power in the SAUs and ICAR Institutes also need regular in service training to upgrade their knowledge and skills in cutting edge technologies.

Besides, a large number of youths and adults lives in rural areas need formal and non-formal training. The rising cost of formal instruction, growing competitions and lack of uniformity of standards are some of the impediments in reaching the poor living in far flung areas. Thus there is ample scope to provide uniform quality instruction, to upgrade skills of teachers, scientists, students and farmers through distance learning.

Organization and Management of Distance Learning

Initiation of distance learning programmes requires intensive planning about courses, target beneficiaries, faculty, infrastructure, counseling and support services, etc. Face–to–face tutorials and student support is another important component in answering many needs of the students. Many distance education institutions have small but highly organized centers for head office, regional centers and study centers close to students. They operate by bringing services closer to students. It is noteworthy to mention that a large number of institutions have interest in distance learning programmes like publishers, booksellers, universities, broadcasters, printers and other.

In each SAUs Directorate of Extension Education is functioning having printing and media facilities such as publication division, video unit etc. One separate cell can be established in which Director of Extension Education will act as nodal officer and will be assisted with one Deputy Director (Distance education) and two Assistant Directors. One of These Assistant Directors will look after technical aspects such as print, electronics media programming, broadcasting etc. Other will see administrative part like finance, administration and management policy formulation.

A number of trends, which include technological developments and the growth of the internet, have accelerated the move to e-learning in educational institutions and the corporate world. Technology and the internet facilitate a more active role in the training process. Distance learning training could be imparted through combination of print, video-cassettes, internet, e-mail, television, radio, telephone, mobile, study centres, counseling, assignments etc. Sreekumar (1998) found that various media employed in distance teaching such as the print media, *i.e.* self-instructional material, audio lessons, video lessons and counseling, facilitated the development of managerial skills. Among these, the print media was found to be more effective in developing the managerial skill as preferred by learners of the management programme. This is followed by other media such as audio, counseling and video. A grouping of media suggested by Gerard (1997), distinguishes media into the following four categories: Print, recorded instruction, real time and computer based. Neuhauser (2002) found in her study of participants enrolled in differing types of online courses. Her qualitative data suggested that highly interactive components such as exercises, activities, animations, and video tended to help engage the student in the learning process, as opposed to more static print and graphic media.

Media Combinations Possible in Distance Learning

1. Print includes self-instructional material (SIM)
2. Student-support includes study centre and tutoring
3. Audio-based technologies includes radio, audio cassettes, telephone, teleconferencing/audio-conferencing, and voice conferencing
4. Video-based technologies includes television and video-cassettes

5. Computer-based technologies includes computer conferencing and E-mail communication
6. Virtual campuses

Explanation about each media combination is given below:

Print Includes Self-Instructional Material (SIM)

In distance education system students are away from the teachers. The teachers' role is performed by SIM which are basically learner-centered materials. Open, distance and flexible learners usually depend a lot on SIM because they have to learn on their own, at a time, pace and place of their own choice. Various forms of support provided in classroom interaction are built into the SIM themselves. SIM differs from textbooks and journal articles which are organized in terms of the subject matter rather than as aids to learning.

SIM are:

- *Self-explanatory*: Learner can understand without external support.
- *Self-contained*: Learner may not need additional materials.
- *Self-motivating*: Materials arouse curiosity and are related to familiar situations.
- *Self-evaluating*: Self assessment questions/exercises, activities, unit-end questions, etc. for providing feedback on performance are provided.
- *Self-directed*: Learner is given necessary guidance, hints, and suggestions at each stage of learning.

Student-Support includes Study Centre and Tutoring

Study centre, provides quality reception and quality tuition to its students, is highly motivating and encourages students to learn. Mills (2000) has described the following support facilities available at study centres:

- A place for individual study in appropriate surroundings and at appropriate times
- Library facilities
- Laboratory and hands-on practical work facilities
- An opportunity for students to meet the distance education institution's administrative staff

- A focus for students' association activities
- A source of information and guidance to the general public in a local area
- Access to technology
- Access to local guidance and counseling, individual and group tuition, group viewing and listening
- Facilities for taking test and examinations.

In some open university like IGNOU about 630 study centres are funded by the institution itself and managed by its full- and part-time staff. Many of these study centres are however, located in the buildings of other colleges. These study centres provide all the functions described above and have a corporate image with which students can easily identify.

Tutoring is course related support provided by an individual for a particular learner or group of learners who also use prepared materials as a resource. It takes various forms but is mainly facilitating the learning of a known group of learners and is not replacing the materials. It brings ODL closer to face-to face teaching of conventional universities. But skills of tutoring required in distance learning are much more complex and sophisticated than those required in conventional system. Distance learning promotes independent learning and active participation of the learner: the two fundamental pillars of education. According to Simpson (2000), a distance tutor has to perform a combination of the following:

- Defining the scope of the course
- Explaining the course
- Developing the study skills
- Assessing the students
- Monitoring progress
- Organizing and supervising hands-on practical work.

Audio-Based Technologies Include Radio, Audio Cassettes, Telephone, Teleconferencing/Audio-Conferencing, and Voice Conferencing

Radio has long been one of the most important forms of audio technology by many higher education institutions. The radio programmes either take up crucial concepts from those covered in the print or clarify them, or they provide an alternative perspective in which to view a concept which is already covered in the printed course material. Radio programmes are not interactive, and they cover only the theoretical part of the course. They do not help in developing hands-on practical skills.

Audio-cassettes have been used by almost all institutions providing some form of distance education. Audio cassettes are much more flexible than radio programmes since they can be listened at their convenience. They also have some limitations.

Telephone has been rarely used in distance learning because it is very expansive medium. Used mainly among students.One of the advantage is that it can be used at any time to talk to anyone having that facilities but it lack visual contact therefore skill can't be transfer through it.

Teleconferencing also known as audio-conferencing mostly popular among developed countries having developed technologies and more distance among tutor and students. Due to latest developments it is now becoming popular in India too.In this mode, tutor sets agenda of the conference and at the other end, which is generally a study centre, a number of students sit across a table with a loudspeaker attached to the telephone.

Voice conferencing is the combination of a telephone and computer to provide a complex voice mail service. By making a telephone call to a number set up by the course provider, students can select from a menu of services provided which can be available even on weekends and holidays.

Video-Based Technologies Include Television and Video-Cassettes

Television has been widely used by many distance learning institutions particularly in developed countries. It has advantage over radio that along with reaching large number of audience it includes visual part as well therefore skill can be transfer through it.

Video-cassettes are a powerful resource in ODL. Like audio-cassette they can be used by students at their convenience. They can also be used to transfer the skill by demonstrating the whole process.

Computer-Based Technologies Include Computer Conferencing and E-mail Communication

Computer conferencing is generally orchestrated by the tutor and involves posting a directional message, like a message on a notice board, which can be accessed by students at their convenience by logging on to a computer either at home or at a study centre.

E-mail communication has also become very common. Students use either their own computers or those at a study centres or an internet café to send e-mail messages to peers or tutors to seek clarifications on any topic of the course.

Student should have access to computers at their homes or nearby computer centers. It is still more convenient if internet connection is available. Internet enabled application of computer allows for student –student and faculty- student interaction. If internet is not available educational material can be provided through diskettes, CD ROMs. However, internet based distance learning is more interactive and effective method. Electronic mail allows rapid correspondence between students and professors. Websites of distance education centers may be provided multimedia lessons with facility for downloading. Use of internet also allows chat discussion and collaboration from multiple locations. In the ninth five year plan Agricultural Research Information System (ARIS) cell were established in each SAU with financial support from ICAR. As a more number of scientists, teachers and students got easy access to computers for preparing database and analysis. Besides, central facility for use of computers by scientists, teachers and students has become possible. Many SAUs have also linked their headquarters with outreach extension centers. This is positive sign. However, with increasing availability of computer centers/cyber cafes in the cities, it is now possible to have computer based distance education. Alternatively commercial cyber cafés with adequate expertise and facilities may be identified and empowered as instructional support centers.

Virtual Campuses

A very recent development in distance learning is virtual university or campus, which provides for all transactions in teaching-learning at a distance online. Students may use the internet to download instructional materials and then interact through e-mail.

Also, they may meet at a fixed time for interaction through video-conferencing. Online libraries and telephone counseling are in place for regular interaction, and the local centres provide for skills assessment, besides e-assessment.

Bates (1996) gave list of criteria on which we need to assess technology before using it:

1. *Access*: how accessible is a particular technology for learners? How flexible is it for a particular target group?
2. *Costs*: what is the cost structure of each technology? What is the unit cost per learner?
3. *Teaching and Learning*: what kinds of learning are needed? What instructional approaches will best meet these needs? What are the best technologies for supporting this teaching and learning?
4. *Interactivity and User-friendliness*: What kind of interaction does this technology enable? How easy is it to use?
5. *Organizational Issues*: What are the organizational requirements, and the barriers to be removed, before this technology can be use successfully? What changes in organization need to be made?
6. *Novelty*: How new is this technology?
7. *Speed*: How quickly can courses be mounted with this technology? How quickly can materials be changed?

IGNOU has initiated a different version of a virtual campus. It offers a Bachelor of Information Technology with options for a Higher National Diploma in Computing and Multimedia. For student support, tele-learning centres have been established using these teaching and support methodologies:

- ☆ Live satellite-based teleconferencing lectures supported through CD/ROM.
- ☆ Recorded video lectures
- ☆ Practical laboratories
- ☆ Computer-based training/tutorials
- ☆ Internet learning resource by internet browsing
- ☆ Online interactive chat and expert through fax, e-mail and telephone (Panda and Chaudhary, 2001).

Development in information and communication technology in the country and their application by the ICAR in SAUs and KVK provide appropriate conditions to think of initiating distance learning programmes. Findings of Rajput and Kumar, 2011 revealed that internet and computer were either provided by most of the KVKs or possessed by majority of trainers. However, the models of instructional design require educators to systematically go through a series of steps from analysis, design, development, implementation, to evaluation.The most crucial aspect is selection of content and media according to learners' needs, preference and the environment. Virtual campuses are useful especially in case of imparting skill in case of training which is mainly skill based. Like IGNOU other universities can also plan for the same for providing distance learning training with the help of virtual campuses.

Infrastructure Utilization

In India at present there are 40 SAUs having Agricultural Information Technology Centers (ATICs) and having linkages with KVKs and research stations. Some of the SAUs have good network of Agricultural Schools or Agricultural Polytechniques. For establishing distance education set up it is economical if the available ATICs, KVKs, Agricultural Schools or Agricultural Polytechniques will be used as study centers. The professional staff of this organization will be the resource persons and it will be the centers of non formal education units which will have direct linkage with SAUs. At present 589 KVKs are spread over the country. Each district probably has one KVK it means that for one district SAUs can have at least one center.

These study centers in each district can be linked up with research stations, Information centers, Panchayats, NGOs, SHGs, and Youth clubs in the cluster of the villages so that it will cover the whole district. These organizations will act as sub study centers and here the stake holders will get all the material and modules.The stake holders can give feedback to the sub study centers or can directly contact the SAU. At present, all the SAUs and KVKs are having internet facility. The study materials can be sent by SAU through internet to all study centers and study centers can download and multiply printing material and CDs as centers requirement. It will save transportation cost as well as immediate delivery will be possible.

Indian Council of Agricultural Research has provided e-connectivity facilities to Krishi Vigyan Kendras (remotely located) and Zonal Project improve e-linkage of KVKs with different stakeholders of agricultural development, including technology generation systems (Agricultural Universities and ICAR Institutes), technology dissemination systems (state and national level extension organizations), technology utilization systems (Farmers organizations, Self Help Groups, etc.), and technology support system (technological input and service delivery organization). This will provide an opportunity to the KVK professionals to interact with researchers, extensionists and farmers in better way and by sharing experiences of each other. Through this facility, the linked centres are able to have access to global e-content on agriculture. KVKs are in the process of developing web pages which include information on district data base on agriculture, technical information, input availability, weather/disease forecasting, market intelligence information, etc.

KVK hub has been established at the ICAR, New Delhi in order to monitor and supervise the activities of KVK. Provisions for two-way audio and one-way video broadcasting facility have been made in the KVK hub. Weekly lectures of agricultural experts are broadcasted through KVK hub which is viewed by all the centres under e-linkage project. The viewers can interact with the expert through two way audio system during the weekly lectures. As KVKs are under supervision of ICAR and SAUs so even if some KVKs are not well equipped with latest IT facilities due to their location we can make use of ICAR institutions and SAUs for providing training through distance mode because ICAR and SAUs are fully equipped with required technologies like computer, internet, even the video conferencing. For providing counseling or any sort of clarification academic institutions are also having sufficient number of trained and knowledgeable teaching staff in different subject's area of concern.

Conclusion

Inclusion of different media combinations in distance learning depends on their availability. In each SAU, Directorate of Extension Education (DoEE) is functioning having printing and media facilities such as publication, video production etc. Regular financial and professional support are provided to SAUs for modernization and strengthening of academic facilities, infrastructure and faculty improvement by ICAR. By utilizing available human resources and infrastructure we can plan for starting distance learning. KVKs could be used as study centres and available academic staff as resource persons. At the TNAU (Tamil Nadu Agriculture University) Coimbatore video conferencing unit has been set up for strengthening the distance learning programme and inter-campus e-learning programme. Most of the above mentioned facilities for study centres are available in KVKs. These study centres can be linked with other units which are already functioning with SAUs like KVKs, Training Centers, NGOs (Non Government Organizations) etc. It will save time as well as cost. By utilizing net facilities available the material can be delivered to the stakeholders without any delay. So after analyzing all the issues it can be said that we have everything available in front of us what all required is to put little more efforts in giving all these a proper shape through required co-ordination.

Section 3
Improving University Teaching

Chapter 13

Professionalizing Teaching in Agricultural Universities: Need and Strategy

Learning is not a spectator sport. Students do not learn much just by sitting in classes listening to teachers, memorizing pre-packaged assignments, and spitting out answers..

– Chickening and Gamson

Introduction

India has developed strong agricultural education system since independence. The establishment of State Agricultural Universities contributed to reorganization and strengthening agricultural education system in the country. The human resource developed in agriculture education system has been instrumental in agricultural transformation in the country. However, the agricultural education system has not been kept pace with the rapid technological development-taking place globally. In the present era of specialization and development in modern science, it is necessary that we restructure our agricultural education in a manner that the graduates coming out are job providers rather than job seekers. The future strategy for agriculture is going to be very different in view of depleting land resources, depleting genetic diversity, over exploitation of natural resources, increased population, increased demand for food and other commodities and lesser chance of employment. The question before us is how to improve our agricultural education and make our graduates competitive nationally and internationally. Of late, there has been widespread consensus that quality of graduates in various disciplines of agricultural sciences does not match with the changing needs. Hence there is urgent need for upgrading the quality of teaching in agriculture. Educational systems are dynamic in nature, continuously adapting to meet the emerging needs of society. In the era of WTO it is very much necessary to improve the quality of teaching for professional development of teachers as well as students.

Education has acquired a new dimension with phenomenal breakthrough in information and communication technologies. The emergence of virtual classroom, open learning systems and distance education have given a new perspective to teaching and learning process. Education during the contemporary times is not confined to classroom teaching alone but it has become life long process according to the needs of the learners. Education technology has evolved out of education to study the problems of education systematically through use of theoretical concepts from physical, social, behavioural sciences and technologies.

Academic improvements do not come by wishes. It takes time, efforts and organizational commitment to improve performance in class rooms. Various stakeholders need to agree on certain norms and procedures for standardizing teaching. Accordingly standards of performance have to be assessed. No doubt, teaching has been given a place of pride in State Agricultural Universities. New teachers definitely need orientation to the standards of appropriate teaching behaviour for which there is little or no preparation prior to joining the University.Thus, most people learn by trial and imitation rather than conscious experimentation and learning.Eventhough, it is generally believed that the agricultural universities are paying equal importance to teaching, research and extension, teaching is less systematic and lesser emphasized in most places. All said there is little corporate planning, supervision and evaluation of teaching in agricultural universities. Once the courses are decided at the department level, it is assumed that everyone will take care of it.

There is no system of monitoring if things go as planned.As a result, teaching is most easy, comfortable and stress free. Internal system of examination has placed lot of liberty on the teachers. Thus, it is not known to what extent the liberty is being used for right cause. Ultimately teachers are promoted and selected for their achievements in research rather than teaching. The reason being, there is no criteria to provide valid evidence for effective teaching. How can we prescribe standards of work performance and their effective evaluation is question of questions.

Teaching Skills Every Teacher Must Know and Use

The teacher has complex task.He should not only know about the subject and learner but also the learning process to create experiences to produce the desired changes in behaviour. The teacher must consider physical and social environment, audio-visual aids available, materials the students will use and the arrangement of the classroom and utilization of pupil activity to create desired experiences. Teaching is not covering syllabus or passing information. Teachers are hired to influence the minds of the learners. It is indeed quite challenging to motivate and enhance learning among students. It is not enough for the teachers to know and understand the subject himself. He has to find ways and means to make the learners know and understand what he has mastered already. This is quite challenging and calls for more than subject matter knowledge. An effective teacher must master the craft of communication. It is said that teaching is not filling the bucket rather it is like lighting the lamp. The phrase emphasizes on the potential of learners. It is learners who will learn. Thus the emphasis should be on the goal of learning. What should learners learn at the end of teaching? How can they learn it during the given time? What approach should I select to achieve my outcome? Learners are not empty pots. They have their own experiences and ideas. A teacher must begin where the learner is and take them to the goal. Human mind is like parachute. It works best when it opens. The teacher must do something to open it. Thus a teacher must learn a number of skills as described below.

Entry Skill

All the verbal and non-verbal communication behaviour of teachers communicates something to the students. Teachers must watch out their entry in the classroom. Do you enter carelessly looking at walls, notes and blackboards or you consciously smile, look at the students and make a few positive remarks. Be conscious and do not forget to look at your clients and greet enthusiastically. This builds positive atmosphere.

How do you being your lecture? Do you start the lesson of the day straight away by writing the topic? The class must be started by appropriately orienting the students to the topic of the day by asking relevant questions. Giving personal experience or interesting case relevant to the topic may catch attention of the students. Alternately, students may be asked to recall the last lesson taught and a link should be established with the topic of the day. The objectives of the lesson should be clearly spelt out and even written on the board to act as road amp for the students. Let students know exactly what is to be learnt, to prepare them for it.

Structured Lesson

Human mind is logical and remembers what is organized logically and in sequence. A teacher must divide his topic into three or four major parts. The parts should be organized in an easily understandable sequence. Each part should be dealt with appropriate introduction, explanation and conclusion so that learners can make sense out of it.

Exposition Skill

Ability to explain requires good vocabulary to explain a difficult term or phenomenon in many alternative ways. Use of examples, evidences and visuals enhance understanding of new concepts.

Handling Media in Teaching

Speaking alone is not enough. What you speak is lost in the air but what you write on the board stays. Plan your board work in advance to put basic essential points on board. Planning visual aids like charts, transparencies or power points beforehand helps to concentrate on explanation. Besides teachers do not have pressure to remember everything. Take care to stand aside and point out the exact point on visual.Use of handouts covering basic minimum content may allow students to concentrate on discussion and problem solving. Besides charts, transparencies, power points and electronic media may be used frequently to support verbal delivery and create variety.

Verbal and Non-Vermal Communication Skill

Speak clearly and loud enough for everyone in the class to hear. Mind your pace of speaking-not too fast, not too slow. In fact, follow the same speed as in normal conversation. Become aware if you are in the habit of repeating some words like I mean, you see, let me tell. Avoid such vocal virus or else you will become a laughing stock..

People perceive messages mostly through non-verbal communication. Position yourself in full view of the students. Look evenly at both sides, Move a little towards students from time to time. Use limited gestures. Use facial expressions to express emotions consistent with the dialogue.

Do students sit passively in your class? Are they taking notes all the time and listen passively to what you speak. It is better to turn the table and ask their inputs. Sometimes, you may ask the class to summarize what you have spoken in ten minutes. Encourage question answer session at the end of the lecture to clarify doubts.

Questioning Skill

Ask different types of questions to know the students' progress. Sometimes you may ask simple question to encourage response by many. You can raise the level of question to know the dept of learning. Direct your question to all the students. Do not always ask a particular group only to respond.

Closure

Every teacher must plan how to end the class. Do not leave the class abruptly. Pull the key points and important explanations together and lead to meaningful conclusion. Tell them appropriate references to be consulted and questions they should think about. You may ask them to come ready for the next lesson

Teaching should be planned and purposeful. A teacher must show positive orientation towards students through verbal and non-verbal communication. Clarity of expression, simple vocabulary, clear explanation, normal pace of speaking and use of appropriate gestures, postures and controlled movement are helpful in conveying meaning.

Roles of Teachers

Planning of Teaching

Each teacher should start preparing for a course well ahead of the semester. To help in advance planning, courses may be allotted toward end of previous semester. In order to motivate teachers for preparation, he may be given choice of courses on the basis of training, experience and specialization. So far there is planning of course at departmental level as a ritual where courses are allotted according to prevailing practices. There is no hard and fast rule about who should be given under-graduate, Master's or doctoral courses. How are Master's and doctoral level courses taught differently than under-graduate courses? It is generally said that elderly and experienced teachers should be given under-graduate courses. Many senior faculty, though interested, may be burdened other extra responsibilities of administration. They may have to skip classes at the last moment for some important meeting with high officials. This may require some unprepared junior to take the classes most of the times. So it is not a question of junior or senior teacher, rather teaching effectively. How can we ensure that planning is done at the level of the teacher and it is adequate? in a reputed business school,the head of the department may not allow the new faculty to teach unless detailed teaching module with lecture details,hand-outs,cases,aids and references is not in place beforehand.Otherwise,it is assumed that the teacher has taken a less than serious approach to teaching

Creative and Standard Course-Outline

Even if the course-syllabus is fixed, course outlines can be revised by including recent terms, relevant topics, references or problems. Preparation of course outline is acreative and time taking task demanding genuine planning skills.In fact preparation of course-outline is taken very lightly. Get an old copy of outline and make multiple copies for distribution. A course outline is a document of learning contract between teacher and student. It should clearly express mutual expectations. Thus, teacher should take time to elaborate contents along the lecture hours, reading assignment writing assignment, list of books to be consulted, standards and type of examinations, probable dates of examination and choice of assignments. Assignments should be decided before hand on the basis of course objectives. The standards of evaluation and deadline must be explained clearly. In fact this should be a self explanatory document for students and other teachers. It must be made available to students and supervisor.

Course Materials

Every teacher should develop different types of media and materials to make classroom teaching effective. Making people learn is not possible through speaking only. Thus, it is not enough to speak in the class. To suplort your verbal presentation use overhead transparencies, flip-charts, samples, computer slides, pictures or simple board. Over a period of time the performance of teachers can be answered on the basis of instructional materials S/he develop gradually, as detailed below:

Hand-out

Simple one tow page hand-outs can be made clarifying important terms, major points with examples and explanation relevant to the lesson. This may help the students to stop writing and start listening with attention. Questions may be built in between and space given for students to jot down. This will help the student to track teacher's presentation and consult selected reference.

Class Notes

All the teachers make class notes for their own consumption. However class notes are made in different ways. It may be in the form of hand written text or key points on reference cards or the whole text systematically organized on transparencies or slide for skillful presentation

Slides/Transperves

It is convenient now to make slides for each and keep a copy for consultation and another for distribution to students. Slides of each lesson can be stored alongwith notes as teaching package. Transperencies of each lesson can be prepared as above and used again and again.

Video

Video programmes relevant to particular lesson can be purchased or produced with the help of media unit to enrich the classroom experience.

Flip-charts

flip-charts are not only easy but less time taking with increasing availability of computers, readable copies of graphics can be produced on chart paper of flex material and preserued.

Teacher's Manual

Manuals are written for trainers to guide them in step-by-step manner to help in conducting training effectively. Similarly experienced teachers may write manuals out of their years of experience of teaching including well developed course outline, smaller versions of transparencies, charts, questions, hand-outs, etc. Such manuals may indicate the professionalism of a teachers.

Use of Cases

Teaching through theortical text book has been conventional approach. Today there is need to inculcate thinking analysis and problem-solving among students. Thus, cases with real-life touch can be good stimulants to encourage discussion and active learning.

Motivation, Planning and Supervision of Teaching

- ☆ Give importance to teaching at college level. Listen to the experiences of teachers. Creat conducive atmosphere to promote good teaching practices by organizing regular discussions and consultations on teaching and learners. Let more and move teacher become aware of basic teaching skills.
- ☆ Recognize good teachers. Create a systematic procedure to identify effective teacher and promote their interest meaning fully. Let some good teachers devote more time towards developing teaching materials and refining teaching skills.
- ☆ Supervise teaching by departmental heads/dean to know content coverage, delivery skills, problems and innovative solutions.
- ☆ Encourage advance planning at the department and college level to discuss finer points, budget, field visit, lab requirement for effective coordination. Invite teachers to present their course plan, teaching materials, evaluation procedures, etc.

- ☆ Irain and motivate new teachers. Orient them to better teaching practices. Let them work together with senior teachers.
- ☆ Encourage team teaching by attractive two or more teachers to plan, implement and evaluate a course together.
- ☆ Good teaching institution should gradually focus more and more on group-discussion, case method, project-work, field visit and actual work experience to provide.
- ☆ Create a cell in the college to encourage research and documentation on teaching. Let some staff collect resources on teaching and disseminate among the interested teachers.

Strategies for Professionalising Teaching

It is a common knowledge that there are wide variations in teaching from college to college, course to course and teacher to teacher. How can we standardize teaching? It calls for efforts to improve teaching by starting regular in-house teachers' training. Presently, there is urgent need to move away from didactic teaching and look for new instructional approaches like student-centered teaching, problem solving, action learning, experiential learning, project mode, etc. It is necessary to experiment with methods of teaching for improvement of quality and professionalism. An educational system is a group of components like students, teachers, media, materials and educational objectives working together in organized fashion. Application of innovations of science and technology along with humanities, social sciences and education is mandatory in achieving the goals of education. It is here that science of educational technologies comes in picture to analyze the process of education and solve the problems of teaching and learning. Thus, advances in educational technologies must now be used to systematically design, implement and evaluate educational process. It is necessary to experiment with methods of teaching for improvement of quality and professionalism.

Continuous Review of Curriculum

Curriculum design is foundation for teaching. There is need to continuously revise curriculum according to the needs of the market. Participatory curriculum design should be practiced to gauze market and bring consensus among stakeholders. The curriculum must focus on development of employable skills, and strategies should be designed to develop those skills. This calls for innovations in teaching methodology.

Preparation of Teaching Package on CD

Normally the contents and methodology developed by individual teachers remain with them and seldom shared. If the teachers retire or shift jobs their unique experiences go away with them. The experience and wisdom of teachers must be preserved in planned manner to incorporate changes and refine courses. Thus every senior teacher may be asked to record their lecture notes along with aids and approaches used by them on computer so as to standardize the teaching objectives, content and methodology. Such packages clearly indicate the course outline, specific behavioural objectives, lecture notes along with photocopies of related reference materials and list of assignments along with criteria of evaluation. The whole strategy can be discussed with colleagues and even shared with experts outside for possible changes. Such teaching packages can be exchanged with teaching faculties in the other universities within the country.

Experiential Learning

Now that professionalism calls for performance and problem solving students must be prepared to perform. This is possible only when students work in simulated environment with similar tools and techniques, which are to be used in real life. This calls for more of hand s on experiences and discussion in classes. Regular interaction with industries and employers to get ready for the desired roles.

Interactive Teaching

Teaching is not about delivering lecture but making learners aware of the concepts and practices in a manner that they get interested to explore the literature and learn on their own. Thus, all the materials to be learnt should be made available to the students in soft or hard copies so that students can come prepared for discussion, analysis and problem solving.

On-line Learning

Teachers must get ready to write teaching packages for distance learners. This will help them learn the ways to prepare on line materials and in turn such materials can be used by face-to-face learners. Besides, students can be encouraged to use on-line materials available on internet.

Center for Educational Technology

In order to enhance the effectiveness of teaching and learning a center for enhancement of teaching and learning, well equipped with support services like video recording, multi- media facility and digital camera should be created in every university on the pattern of North American universities. Facility for video recording of teaching to interested teachers may be available on demand. Teachers may be provided feedback and counseling by the concern specialists. Training of teachers and students will be organized regularly on teaching and learning. The center will collect feed back of students and teachers on new practices. It may also help in formation of quality circle and mentoring besides help in preparing teaching packages in digital form and teachers' portfolio will be provided

Team Teaching

It refers to sharing of responsibilities of instruction between two or more individuals. All the team members are involved in planning, implementation and evaluation of teaching. Both the teachers remain together and share responsibilities of instruction. Diverse talents of teachers can be utilized for the benefit of the students. Teachers may provide their observations to each other for improvement. Working in co-operation with each other may create motivation for the new teachers. New teachers may learn by doing together with senior teachers. Even in case of absence of one of the teachers, the continuity of the course is not affected adversely. Working together and taking turns in teaching may provide variety and overcome boredom due to monologue

Chapter 14

Learning Theories and their Implications in Teaching

Teaching and learning are complimentary to each other. The desirable change in behaviour is brought out by teaching and learning. The word 'learn' comes from Middle English *"Lernen"* meaning to learn or to teach. *Lernen* originated from 'learn' meant learning or teaching but now used to mean that which is being taught (Oxford University Dictionary 1955).

The word 'Teach' comes from Latin word *'Taecan'* meaning to teach. To teach means to show a person how to do something. In generic sense teaching denotes action undertaken with intension of bringing about learning in another.

Learning is defined as relatively permanent change in behaviour that occurs as a result of practice or experience.

According to John Dewey teaching entails learning. Dewey stated that teaching and learning are interrelated concepts and viewed as what selling is to buying, since there is no selling, there is no buying. If there is no teaching there is no learning. Teaching is both art and science. The effective teaching requires deep understanding of the scientific principles and theories guiding learning process.

Learning theories are broadly grouped under three schools of thoughts these are:

- ☆ Behaviourism
- ☆ Cognitivism
- ☆ Constructivism

Behaviourism

This school of thought is based on observable changes in behaviour. Behaviourism focuses on a new behavioural pattern being repeated until it becomes automatic. The theories which are grouped under behaviourism interpret learning in terms of stimulus and response. Behaviourism can be traced back to Aristotle, whose essay "memory" focused on association being made between events such as lighting and thunder. The theory of behaviourism concentrates on the study of overt behavour that can be observed and measured (Good and Brophy, 1990). It views that response to stimulus can be observed quantitatively totally ignoring the possibility of thought processes occurring in the mind. Some key players in the behaviourist theory were Pavlov, Watson, Thorndike and Skinner.

Pavlov's Classical Conditioning

This is the simplest form of learning in which a neutral stimulus (Conditioned Stimulus) after repeated pairing with an unconditioned stimulus, elicits conditioned response which was earlier emitted by unconditioned stimulus.

Pavlov's Experiment: Pavlov was Russian physiologist who conducted experiments on dog. Food (Conditioned stimulus) produced saliva (Unconditioned response) after pairing food (US) with bell (CS) a number of times, conditioned response (Saliva) was emitted by conditioned stimulus as given below:

Food	Unconditioned Stimulus
Salivation	Unconditioned Response (natural, not learned)
Bell	Conditioned Stimulus
Salivation	Conditioned Response (to bell)

Essentials of Conditioning

- ☆ Number of pairings

- ☆ Contiguity
- ☆ Inter-stimulus interval
- ☆ Reinforcement
- ☆ Extinction
- ☆ Reconditioning

Thorndike's Trial and Error Theory (1874–1949)

Edward Thorndike did research on animal behavior before becoming interested in human psychology. He set out to apply "the methods of exact science" to educational problems by emphasizing "accurate quantitative treatment of information". "Anything that exists, exists in a certain quantity and can be measured" (Johcich, as cited in Rizo, 1991). His theory, Connectionism, stated that learning was the formation of a connection between stimulus and response.

Experiment

A hungry cat was confined in a box which could be opened by pulling a lever. A plate of fish was kept outside the cage visible to cat. After making haphazard movements she succeeded in opening the door. Through repetitions of the experiment, cat showed gradual improvement and finally learned to open the door.

Essentials of Learning

- ☆ Drive
- ☆ Goal
- ☆ Block
- ☆ Random Movements
- ☆ Chance Success
- ☆ Selection of right responses
- ☆ Fixation

Application of Thorndike Theory

- ☆ The **"law of effect"** stated that when a connection between a stimulus and response is positively rewarded it will be strengthened and when it is negatively rewarded it will be weakened. Thorndike later revised this "law" when he found that negative reward, (punishment) did not necessarily weaken bonds, and that some seemingly pleasurable consequences do not necessarily motivate performance.

- ☆ The **"law of exercise"** held that the more an S-R (stimulus response) bond is practiced the stronger it will become. As with the law of effect, the law of exercise also had to be updated when Thorndike found that practice without feedback does not necessarily enhance performance.
- ☆ The **"law of readiness"** because of the structure of the nervous system, certain conduction units, in a given situation, are more predisposed to conduct than others.

Thorndike's laws were based on the stimulus-response hypothesis. He believed that a neural bond would be established between the stimulus and response when the response was positive. Learning takes place when the bonds are formed into patterns of behavior (Saettler, 1990).

Watson (1878–1958)

John B. Watson was the first American psychologist to use Pavlov's ideas. Like Thorndike, he was originally involved in animal research, but later became involved in the study of human behavior.

Watson believed that humans are born with a few reflexes and the emotional reactions of love and rage. All other behavior is established through stimulus-response associations through conditioning.

Watson's Experiment

Watson demonstrated classical conditioning in an experiment involving a young child (Albert) and a white rat. Originally, Albert was unafraid of the rat; but Watson created a sudden loud noise whenever Albert touched the rat. Because Albert was frightened by the loud noise, he soon became conditioned to fear and avoid the rat. The fear was generalized to other small animals. Watson then "extinguished" the fear by presenting the rat without the loud noise. Some accounts of the study suggest that the conditioned fear was more powerful and permanent than it really was. (Harris, 1979; Samelson, 1980, in Brophy, 1990)

Certainly Watson's research methods would be questioned today; however, his work did demonstrate the role of conditioning in the development of emotional responses to certain stimuli. This may explain certain fears, phobias and prejudices that people develop.

Skinner's Operant Conditioning Theory (1904–1990)

The theory has been based on Thorndike's learning by selecting and connecting under the law of effect. Most of S-R theorists assume that stimulus is a necessary prerequisite for evoking a response.

He opposed S-R formula and proposed R-S formula.

Difference Between Classical and Operant Conditioning

Classical Conditioning (Pavlov)	Operant (Instrumental) Conditioning (Skinner)
Unconditioned Stimulus ——> Unconditioned Response (food) (salivation)	Response ——> Stimulus (reward) (press lever) (food)
Unconditioned Stimulus ——> Unconditioned Response (food) (salivation) Conditioned Stimulus / (bell)	TIME
Conditioned Stimulus ——> Conditioned Response (bell) (salivation)	Conditioned Response ——> ConditionedStimulus (press lever) (Reward)(food)
In classical conditioning, a neutral stimulus becomes associated with a reflex. The bell, a neutral stimulus, becomes associated with the reflex of salivation.	In operant conditioning, the learner "operates" on the environment and receives a reward for certain behavior (operations). Eventually the bond between the operation (pressing the lever) and the reward stimulus (food) is established.

When desired response is emitted a reinforcing stimulus is presented. So the response is conditioned by reinforcement. He proposed four types of reinforcement:

1. *Positive Reinforcement or reward*: Responses that are rewarded are likely to be repeated. (Good grades reinforce careful study.)
2. *Negative Reinforcement*: Responses that allow escape from painful or undesirable situations are likely to be repeated. (Being excused from writing a final because of good term work.)
3. *Extinction or Non-Reinforcement*: Responses that are not reinforced are not likely to be repeated. (Ignoring student misbehavior should extinguish that behavior.)

4. *Punishment*: Responses that bring painful or undesirable consequences will be suppressed, but may reappear if reinforcement contingencies change. (Penalizing late students by withdrawing privileges should stop their lateness.)

Application of Operant Conditioning

- Various forms of reinforcement should be used.
- Care should be taken for implementing proper schedule for reinforcement.
- Applicable in shaping and modification of behaviour.
- External forces in the environment must maximize the satisfaction of learner.
- Programmed instructions can be used for preparing learning materials.

What is Cognitivism?

"Cognitive theorists recognize that much learning involves associations established through contiguity and repetition. They also acknowledge the importance of reinforcement, although they stress its role in providing feedback about the correctness of responses over its role as a motivator. However, even while accepting such behavioristic concepts, cognitive theorists view learning as involving the acquisition or reorganization of the cognitive structures through which humans process and store information." (Good and Brophy, 1990, pp. 187).

As with behaviorism, cognitive psychology can be traced back to the ancient Greeks, Plato and Aristotle. The cognitive revolution became evident in American psychology during the 1950's (Saettler, 1990). One of the major players in the development of cognitivism is Jean Piaget, who developed the major aspects of his theory as early as the 1920's. Piaget's ideas did not impact North America until the 1960's after Miller and Bruner founded the Harvard Center for Cognitive studies.

Tolman's Sign Learning Theory: it is an eclectic approach-combines advantages of S-R and cognitive theories. Tolman strongly opposed S-R associationism but his system is rooted in behaviourism and is known as purposive behaviourism. He highlighted the role of intervening variables which are not directly observable. It is a molar approach to study behaviour which is goal seeking and purposive behaviour is a function of various variables given by a formula:

$$B = F\,(S.P.H.T.A.)$$

where,

S: Environmental Stimuli

P: Physiological Drives

H: Heredity

T: Previous Training

A: Maturity of Age

Learning is cognitive mapping to find new solutions. Reward expectancy affects learning. Dormant learning occurs in behaviour and comes to surface when required.

Application of Tolman's Theory

- ☆ Set clear-cut goals/Objectives
- ☆ No attempt or no learning is wasteful, no immediate purpose is served when need arises
- ☆ Capacity of learner should be taken care of.
- ☆ Reward and punishment is not essential at each step. Emphasize intrinsic motivation.

Kohler's Insight Theory of Learning/Gestalt Theory

This theory was a revolt against the prevailing S-R approaches to learning that complex human behaviour can not be the mere connections but based on cognitive power or perceptual organization, insight. It is a purposive, imaginative and creative enterprise.

Experiments- Series of experiments was consulted on Chimpanzees. Chimpanzees were able to get bananas hanged from the ceiling of a cage (*i*) by using a stick, (*ii*) by joining two sticks, (*iii*) by using a box and (*iv*) by placing two boxes over each other. Learning was the result of insightful behaviour of Chimpanzees and ability to see the relationship between different elements of situation.

Insight Depends Upon

- Capacity of the learner
- Past experiences
- Survey of the problematic situation
- Attitude of concentration
- Trial of response
- Sudden solution

Application of Kohler's Theory

- Capacity of learner *i.e.* basic intelligence should be taken into consideration.
- Previous knowledge and experience should be kept in mind.
- Learning should be goal oriented and purposeful. Learner should be well acquainted with goal.
- Interest and curiosity of learner should be aroused.
- Structure and organization of subject matter should be kept in mind rather than isolated facts.
- All aspects of situation should be taken care of *i.e.* problem should be considered as a whole.

The Basics of Constructivism

Bartlett (1932) pioneered what became the constructivist approach (Good and Brophy, 1990). Constructivists believe that "learners construct their own reality or at least interpret it based upon their perceptions of experiences, so an individual's knowledge is a function of one's prior experiences, mental structures, and beliefs that are used to interpret objects and events." "What someone knows is grounded in perception of the physical and social experiences which are comprehended by the mind." (Jonasson, 1991).

If each person has his own view about reality, then how can we as a society communicate and/or coexist? Jonassen, addressing this issue in his article *Thinking Technology: Toward a Constructivist Design Model*, makes the following comments:

- "Perhaps the most common misconception of constructivism is the inference that we each therefore construct a unique reality that reality is only in the mind of the knower, which will doubtlessly lead to intellectual anarchy."

- "A reasonable response to that criticism is the Gibsonian perspective that contends that there exist a physical world that is subject to physical laws that we all know in pretty much the same way because those physical laws are perceivable by humans in pretty much the same way."
- "Constructivists also believe that much of reality is shared through a process of social negotiation..."

If one searches through many philosophical and psychological theories of the past, the threads of constructivism may be found in the writing of such people as Bruner, Ulrick, Neiser, Goodman, Kant, Kuhn, Dewey and Habermas. The most profound influence was Jean Piaget's work which was interpreted and extended by von Glasserfield (Smorgansbord, 1997).

Realistic vs. Radical Construction

Realistic constructivism–cognition is the process by which learners eventually construct mental structures that correspond to or match external structures located in the environment.

Radical constructivism–cognition serves to organize the learners experiential world rather than to discover ontological reality

(Cobb, 1996, in Smorgansbord, 1997).

The Assumptions of Constructivism–Merrill

- Knowledge is constructed from experience
- Learning is a personal interpretation of the world
- Learning is an active process in which meaning is developed on the basis of experience
- Conceptual growth comes from the negotiation of meaning, the sharing of multiple perspectives and the changing of our internal representations through collaborative learning
- Learning should be situated in realistic settings; testing should be integrated with the task and not a separate activity.

Thus the important conclusion drawn based on these three type of theories can be summarized as follows. Behaviourism studies obvious behaviours. It include behaviours that can be observed and calculated. Behaviourism has created a wide impact on different educational areas such as systems approach; computer assisted learning, objectives etc. Cognitivism deals with internal mental processes of the mind and how these processes could be used to endorse effective learning. In cognitivism the task is analyzed then broken down into small pieces. The information is than taught from simple to complex. The constructivism on the other hand promotes the a learning experience where the learning approach for each learner would be different where the methods and results cannot be easily measured.

http://www.usask.ca/education/coursework/802papers/mergel/brenda.htm

http://www.outsource2india.com/LearningSolutions/articles/learning-theories.asp

http://en.wikipedia.org/wiki/Learning_theory_(education)

Chapter 15
Skills in Teaching

Teaching is not merely covering syllabus or passing information. Teachers are hired to influence the minds of the learners. It is indeed quite challenging to enhance learning among students. It is not enough for the teachers to know and understand the subject. They have to find ways and means to make the learner know and understand what they have mastered already. This is quite difficult and calls for more than subject matter knowledge. An effective teacher must master the craft of communication. It is learners who have to learn. How can they learn it during the given time? What approach should be selected to achieve learning outcome? Learners are not empty pots. They have their own experiences and ideas. A teacher must begin where the learner is and take them to the goal. Human mind is like parachute. It works best when it opens. The teacher must do something to open it. Thus, a teacher must learn a number of skills for teaching effectively.

Entry Behaviour

All the verbal and non-verbal commutation behaviour of teachers communicate something to the students. Teachers must watch out their entry in the classroom. Do you enter carelessly looking at walls, notes and blackboards? Do you consciously smile, look at the students and make a few positive remarks. Be conscious and do not forget to look at students and greet them enthusiastically. This builds positive atmosphere.

Opening Remarks

Students come to your class from hostel or a last class with entirely different subject or after an hourly examination. Take time to draw their attention towards the lesson of your class. You may ask one or more students to recapitulate the gist of the last class. You may ask relevant questions. Giving personal experience and interesting cases relevant to the topic may catch attention of the students. Purpose of the lesson should be clearly spelt out and even written on the board to act as roadmap for the students. Let students know exactly what is to be learnt, to prepare them for it. Various techniques to begin the class may be:

- ☆ Recall the last class.
- ☆ Stress importance of today's lesson.
- ☆ Share a case/personal experience related with the lesson.
- ☆ Ask them to share something they have known or experienced about the topic.
- ☆ Discuss a current news item related with topic.

Students are not able to understand the major theme, if teacher has not clearly specified the purpose. So let them know before hand what four or five things they are going to learn today, give them a little overview to create interest. This will make students attentive and alert about what is to be covered.

Designing Lesson

An old German maxim states,

"All that is said is not listened

All that is listened is not understood

All that is understood is not accepted

All that is accepted is not done"

So there is a large gap between what they hear, understand and do. So the impact of lecture is less in spite of all the efforts. People understand things better that are logically organized. Logical organization demands organization from simple to complex, empirical to rational, concrete to abstract, and know to unknown. A well-designed lecture should consider the obstacles in communication from learners' point of view. As a seasoned teacher, you may know their levels of understanding. Ability to organize subject matter logically and in sequence is essential.

- ☆ A teacher must divide his topic into three or four major parts.
- ☆ Each part should have appropriate introduction, explanation and conclusion so that learners can make sense of it.

Example

To teach Rural Development Programmes of India, the initial class demands logical organization by making the learners convinced about relevance of studying development programme, concept and functions of different agencies. Each part can be developed separately.

1. Why studying rural development is important for agricultural graduates ?
2. What does rural development mean and how does the government look after it ?
3. How does different organization/agencies get collectively involved in implementation of rural development programmes?

Explanation

Explaining the content is a complex task. Some teachers are able to explain effectively because of rich fund of examples and experiences. Understanding the level of students is key to explanation.

Difficulty in understanding arises due to lack of pre-requisite knowledge.

- ☆ Did you explain the new terms?
- ☆ Did you make sure that students know the background information?
- ☆ Was the language easy to follow.

If the answer to these questions is yes, then what strategy do you have to expand the content ? Ability to explain requires description of difficult terms or phenomenon in many alternative ways. It also demands use of examples, evidences and visuals to enhance understanding of new concepts. You must relate with something already known by students. Besides, summarize points from time to time so that students remember what is being covered.

Use of Audiovisual Aids

Speaking alone is not enough. What you speak is lost in the air but what you write on the board stays. Plan your board work in advance to put basic essential points on board. Planning visual aids like charts, transparencies or power points beforehand gives time to explain. Besides, teachers do not have pressure to remember everything. Take care to stand aside. Speak slowly and point out at the visual.

Use of Verbal Communication

Speak clearly and loud enough for everyone in the class to hear. Pace of speaking should not be too fast or too slow. In fact, follow the same speed as in normal conversation. Become conscious if you are in the habit of repeating some words like 'I mean', 'you see', 'let me tell'. Avoid such vocal virus or else you will become a laughing stock.

Use of Non-Verbal Communication

People perceive message mostly through non-verbal communication. Position yourself in full view of the students. Look evenly at both side. Move a little towards students from time to time. Use limited gestures. Use facial expressions to express emotions consistent with the dialogue. Your posture and facial expressions communicate your real intention.

Student's Participation

Do students sit passively in your class? Are they taking notes all the time and listen passively to what you speak? It is better to turn the table and invite their inputs. Sometimes, you may ask the class to summarize what you have spoken in ten minutes. Encourage question answer session at the end of the lecture to clarify doubts.

Use of Questions

Questions are important stimulant for learners. Questions make them think. Many different types of questions can be asked for different purposes, as below:

To Check Awareness

Easy questions can be asked in between lecture to check alertness/deftness of the students. Many students may volunteer to respond and thus, a positive atmosphere is created. It breaks boredom of one way communication.

To Test Knowledge

Carefully worded questions may be raised to check understanding.

To Help in Application

Practical problems may be given to solve by using relevant theory.

To Develop Critical Thinking Ability

Question of high order may check analytical ability of students.

Thus, questions may vary from low level to high level depending upon the need. However, it is also important to determine who should be asked.

Ask to the Class

Address the question in general to the whole class to see how many people volumes to speak.

Ask to a Group

Question may be addressed to a group at the back, front or side who may be engaged in side-conversation or other diversion.

Ask a Person

Question may be addressed to a person to check his, her performance in particular. Normally questions are addressed to the class.

Handling Students' Questions

Students seldom raise questions but when they do this, must be attended to properly. Student's questions are prizes to the teachers. They indicate that the student is attentive and involved. Teacher may return the question to the class to see if someone knows already. He may rephrase the questions and ask class for answer. He may even answer himself, though it is not necessary to respond himself. In case a teacher does not know the answer, he should admit ignorance and promise to answer in the next class.

Handling Students' Response

Listening is key to responding the response. The students should be listened to carefully. He/she should be complemented for the part of the response which is right. Student should be given correct answer with explanation. Thus, questioning and handling students' responses are important skill to be used purposively.

Ask different types of questions to know the students' progress. Sometimes you may ask simple question to encourage response; by many. You can raise the level of question to know the depth of learning. Direct your question to all the students. Do not always ask a particular group only to respond.

Closure

Do not leave the class abruptly, like the opening, closure has a purpose. It must be planned and linked to the overall lecture. Unfortunately many teachers stop abruptly at the end of fifty minutes or stretch till the next teacher knocks. The activities to be performed at the close of the lecture are as follows:

Pull Out Ideas from the Lecture

Ask students to recall what has been presented today. Help them in recalling important materials.

Help in Application of the Material

It is desirable that students in higher education get to know the practical implications of the materials covered in the class. Problems related with the topic may be discussed.

Achievement of Objectives

It is important to describe the content discussed in the light of the objectives. Students must understand what has been accomplished in the class.

Forward Planning

In order to prepare students for the next assignment to be completed., you may give a preview of next class and ask students to bring some observations to get them ready. In a nutshell,

- ☆ Pull the key points and important explanations together and lead to meaningful conclusion.
- ☆ Ask them to recall key points.

- ☆ Tell them the appropriate reference to be consulted and question they think about. You may ask them to come ready for the next lesson.
- ☆ If possible tell them about the next class.

Conclusion

Teaching should be planned and purposeful. A teacher must show positive orientation towards students through verbal and non-verbal communication. Clarity of expression, simple postures and controlled movement are helpful in conveying meaning. Above all, the way you move in step by step manner indicates your adeptness at teaching skill. Summarize from time to time and remind the key points. Last two -three minutes should be spent on leading the class to conclusion.

Brain vs Mind

The ultimate aim of education is to enable individuals to become the architects of their own education and through that process to continually reinvent themselves. I start with the assumption that in a certain significant sense, mind is not present at birth. Minds are invented when humans interact with the culture in and through which they live. Brains are biological. They are conferred at life's beginnings. Minds are cultural; and although there is not sharp line between what is biological and what is cultural–they define each other–the overriding perspective I want to commend is that schools have something significant to do with the invention of mind. The invention of mind in schools is promoted both by the opportunities located in the curriculum and by the school's wider culture. They are found in the forms of mediation through which the curriculum and schooling as a culture take place. In this sense, the curriculum is...a mind-altering device.

To one who, journeying through night and fog,
Is mired neck-deep in an unwholesome bog,
Experience, like the rising of the dawn,
Reveals the path that he should not have gone.

–Joel Frad Bink

Chapter 16

Methods of Teaching

Method of teaching is important from the point of delivery of subject matter. Even though lectures are most extensively used in higher education, there is need to understand other methods as well to make teaching interesting and interactive. All of us recall how we used to be restless in classroom lectures because of one or other reasons. There is reason enough to combine different methods to achieve different objectives of learning. Of various methods lectures, group discussion, demonstration, buzz session, field visit,role play,case study, brainstorming,practical sessions or field learning etc. selection of method depends on the objectives, group size, time in hand,resources and other considerations. However, we may learn and try various methods to make our teaching effective.

Lecturing

Lecture is probably one of the oldest and most common methods of instruction. Ease in use and economy dictate its wide spread use in all classrooms. However, the fact remains that it is one-way communication. Students sit passively listening or taking notes and are often carried away by the internal thoughts. There are a many barriers between teachers and students. Teachers' voice, pronunciation, pace of delivery, use of black/white board and style have lot of impact on learning

Advantages

It is god to cover large groups of audience. It is easy to use compared to other teaching methods. Within a short time large chunk of material can be taught.It can be used just almost everywhereLecture allows a large volume of materials to be covered in a relatively short period of time. This explains its proven appeal to situations when time considerations are over-riding (as in intensive training courses). The effectiveness of) lecture can be boosted if it is accompanied by appropriate visual support and written material

Disadvantages

Lecture does not address the needs of the individual learner; nor does it provide opportunities for the audience to participate Besides, most people are unable to concentrate for more than a relatively short period. Thus assimilation tends to be haphazard and selective. Parts of the material may even be missed altogether or inaccurately recorded. Moreover, in the absence of feedback it is difficult to judge whether or how much of the materials covered is actually taken in. Retention rates are quite low, unless opportunities are provided for discussion, practice pr simulation. Most of what is 'heard' in the lecture room typically tends to be rapidly forgotten.

Etiquettes for Lecture Delivery

A lecturer is a performer in the classroom. He is manager of the session.How should he behave to promote learning.

- ☆ A teacher should dress appropriately to projecta professional look.Arrive at the door little before time.Greet students cheerfully and start on time.
- ☆ Take time to introduce the topic and create interest. Explain session objectives.
- ☆ Involve audience by using buzz group, group discussion, brainstorming or any other interactive method.
- ☆ Clarify difficult terms and use examples to explain points.
- ☆ Keep voice sufficiently loud and clear. Make sure that pronunciation is clear and well understood.
- ☆ Move at a slower pace in the beginning. Gestures can be used to reinforce learning process. Use visual aids. Stand aside while using visual aids.

- ☆ Use question and answer sessions to clarify doubts.
- ☆ Encourage group interaction
- ☆ Maintain eye contact. Face the learners most of the time
- ☆ Enter the classroom with definite plan Move in the class in planned way to observe all the students.
- ☆ Write gist/key points on chart/board.
- ☆ Give plenty of examples.
- ☆ Explain according to level of learning.
- ☆ Ask questions from time to time. Monitor students' reactions.
- ☆ Use planned closing by summarizing the important points and their relevance suitable reading assignment and topic of next class Try to finish on time.
- ☆ Stand rather than sit during teaching.

Lesson Plan Preparation for Effective Lecturing

A good lecture or class must be planned to use time and recourses effectively. It is better to teach systematically based on understanding and experiences about learners and subject. Lesson plan helps teacher in professional manner. There is nothing as fatal as unpreparedness of a teacher.

What is Lesson Plan?

- ☆ Lesson plan involves defining objectives, selecting and arranging the subject matter and determining the method of procedure.
- ☆ Lesson plan is the title given to a statement of the achievements to be realized and the specific means by which these are to be attained as a result of the activities engaged during the period.

There must be sensible build-up in the course. Lesson must fit in a sequence and each lesson itself must include various subjects. The building blocks for a lesson are acquired by formulating answers to the following questions:

What are the Objectives of the Lessons?

Decide the purpose of the lesson. What would be the outcome? It indicates what the students be able to do at the end of the lecture.

- List 4.
- I ways to ensure a good atmosphere in the group?
- Users base-line situation of the audience?
- Select appropriate methods of teaching.
- Explain one method to assess audience reaction.

Significance of Lesson Planning

- It helps in organizing subject matter suitably: A teacher gets opportunity to arrange subject matter in order to make learners understand. It helps in recollecting appropriate examples and explanation to drive points home.
- It helps teacher to rehearse and prepare for the class beforehand by writing down the points before hand, teacher can rehearse and prepare to be natural in an actual class. He can overcome so me mistakes through practice.
- It helps to monitor and evaluate the class if lesson plans are prepared, someone can observe and assess if steps were followed as desired.
- It helps in training new teachers those with title or no experience in teaching, can prepare and organize their teacher.
- It avoids in meaningless repetition of contents or topic.
- It helps in making check list to consider during teaching: A teacher can consciously remember things to be done during the class by notices it down on the plan.

Lesson Plan

Topic

- Course No.
- Date
- Venue
- Duration

- ☆ Type of students
- ☆ Number of students
- ☆ Starting situation
- ☆ Objectives
- ☆ Time
- ☆ Brief content
- ☆ Didactical Methods
- ☆ Teaching aids

Team Teaching

It refers to sharing of responsibilities of instruction between two or more individuals. All the team members are involved in paining, implementation and evaluation of teaching. Unlike conventional methods of teaching where teachers may divide part (s) of the course to be taught in team teaching both the teachers remain together and share responsibilities of instruction together.

Rationale

Many teachers feel uncomfortable in working together and resist team teaching. The complex task of teaching does not consist of lecturing alone. A teacher needs to carry out discussion or use media or control large number of students within limited time.Thus, it is helpful and harmonious to work in team and implement creative ideas together. This also helps interested younger teachers to learn by seeing and doing together.

Advantages of Team Teaching

1. Diverse talents of teachers can be utilized for the benefit of the students. A teacher may be experienced in use of teaching or may have ability of producing instructional aids. Another teacher may have good command over subject. Thus, the combination may have better impact on learning.
2. Teachers may provide their observations to each other for improvement.
3. Working in co-operation with each other may create motivation for the new teachers. New teachers may learn by doing together with senior teachers.

4. Even in case of absence of one of the teachers, the continuity of the course is not affected adversely.
5. Working together and taking turns in teaching may provide variety to students
6. It helps in appropriate hands-on training of the junior teachers.
7. It is possible to carry out evaluation and research on one's own teaching.
8. It is possible to use media, active learning approaches and innovations in teaching.

Limitations

It is difficult to form a team by official order. People may resist working together or feel the barriors of position or criticism and think it a waste of time.Thus,the method can not be forced but can be initiated with the help of volunteers.

Creating Active Learning Situation in Classroom

An old proverb says I hear, I forget. I second I remember. I do and I understand" people learn better when they are active and involved. Doing thing things and reflecting experience can facilitate learning. A learner may read, discuss, solve problem, present, play; roles etc. learning is not passive. Thus, a teacher may do a number of things to encourage active learning.

- ☆ Let student write something in the mid of a lectures, student may be asked to write four examples of the process being discussed. This will break monitory, make students active and help them think.
- ☆ Let them come to board to explain a point. Calling students to explain breaks one way communication. Other students also start thinking.
- ☆ Let students teach a small portion of the them. Students often become much more responsible and prepare well to face their own classmates.
- ☆ Ask questions that help students think about practical issues. This may provide new inputs and broaden learning.

- Give them a problem to solve. Courses having practical implications may be taught in manner that after every major theoretical lesson, some problems may be posed to students to think about solutions on the basis of concept.
- Let students discuss among themselves in between a lecture. After thirty minutes of class, a discussion may be organized for ten minutes to summarize what has been taught.
- Encourage students to bring news items, examples and quotations related with the course.
- Give small non-gradual but interesting quizzes to get feedback on their learning.
- Invite students to ask questions and raise doubts.
- Conduct debate in the class on controversial issues.
- Prepare interactive hand outwith questions and space far writing so that students can jot down their points, when asked.

Buzz Group

This technique is a modified form of brainstorming, which aims to generate discussion generate new ideas and solutions.

A buzz group normally consists of about 5-6 members. The group may be asked to produce ideas on a narrow or an open-ended topic within a specified time limit. It is customary for the group to appoint a reporter to convey their findings at the end of their deliberations.

The range of topics assigned for buzz group discussion may vary widely. Some examples are given below:

- Thinking about applications of a theory
- Raising questions on a topic
- Drawing a list of issues
- Identifying the causes of a particular problem.

This approach can also be used occasionally in the lecture room, where the lecturer stops for a few minutes and asks small groups of people sitting next to each other to deliberate on a particular issue. One advantage of this approach–apart from breaking the monotony of the lecture and injecting a sense of participation–is to check learners' understanding of particular issues before proceeding to other, more complicated, ones.

Buzz group, however, differs from brainstorming in two respects:

- ✰ Smaller, more informal nature: as stated above, a buzz group consists of a small group of participants 'buzzing' together for a relatively short period of time (sometimes a few minutes).
- ✰ Discussions: contrary to brainstorming, where criticism is deferred until after ideas have been pulled together, in the buzz group, discussion features an important part of the process of in-group exchange of ideas.

Case Method

A case is detailed description of lifelike situations based on adequate data throwing light on the problem. It is quite popular in management training. It provides realistic data to initiate intense discussion, analysis and decision-making.

Case study involves detailed examination of a specific situation. The situation or experience may be real or imaginary but it must bring out facts necessary to make objective decisions. It should be relevant to the purpose of training. It develops skills to analyse situation critically and apply relevant principles. It develops analytical skills. The classroom becomes lively and participative. Unlike lectures, the participants are not at the receiving end of learning but active partners in generating knowledge. The active co-operation of participants is possible only when they are motivated and alert.

Steps in Using Case Method

Problem Diagnosis

A case is introduced by the trainer explaining the goal.The trainees may then go through the case individually or in group to understand the problem. Understanding the problem calls for acquaintance with the actual situation to gain perspective. The task is to identify different components of the problem. This requires a detailed scrutiny of the case so as to understand its features.

Possible Solutions

Identifying problems inevitably lead to the need to formulate possible solutions. Alternative courses of action need to be explored to weigh their consequences. In this way, the effect of each path may be considered objectively to decide the best solution.

Analysis and Conclusion

The alternatives suggested as feasible by individuals and groups may be analysede critically to reach a consensus. Besides, other general issues are raised by the case may be discussed to broaden the horizen.? Often cases can be useful springboards into more general discussions of principles and policy

Advantages

1. Case method brings a sense of realism about the situation through detailed description. This helps the participants to assume the responsibility for problem solving.
2. Case method keeps the participants active through equal contribution and intensive interaction. Thus, trainer is not the sole expert. People learn from each other
3. It helps in developing systematic approach to problem solving.

Limitations

1. Cases are difficult to write. Readymade cases are not available in all the disciplines.
2. Expert resource persons are needed to conduct discussion of cases critically to write appropriate case studies.
3. Training through case study takes lot of time.
4. Cases may not fit all typesf learning objective.

Group Discussion

Discussion refers to verbal interaction between two or more people.

Discussion is preferred as a method of teaching because it suits the characteristic of adult learners. Adults have experiences, which need to be shared, explored and analysed to expand their knowledge. Discussion provides them an opportunity to participate in the process of learning. This is an ideal method because the control rests in the hands of learners. Trainers act as coordinators or facilitators. How ever, the role is very challenging. It requires special skills to create a conducive environment so that learners feel relaxed and motivated to contribute openly.

Strong Points of the Method

It Helps in Exchanging Experiences Among Learners

Trainers and teachers know it very well that lectures make learners passive. They become silent. It is very difficult to know if they are learning or not. Sharing means their experiences are valuable. This boosts their confidence and helps them open up.

It Uses Collective Wisdom of the Group of Learners

It is indeed very helpful to get a variety of ideas on the subject of discussion. Diversity of ideas enriches the process of learning. It expands their horizon and provide new perspective. Both trainers and learners learn together. It is not limited to the knowledge and experience of trainer alone.

It Involves Learners Fully in the Process

Learners are active and in control of the process through out. The trainer has to create conducive physical and psychological climate so that discussion can begin in the right note. The trainer does not interfere in the discussion process. His role is one of an observer and facilitator in analyzing the findings or reports of group members.

It Helps in Improving Communication and Leadership Skills of the Learners

Group discussion is a spontaneous activity where the participants reveal their natural behaviour. They cannot rehearse and act. Continuous exposures to group situation help them learn ways to express themselves. They learn to listen to other people's ideas, analyze information and respect viewpoints different than their own.

Weak points/Limitations

1. As all participants do not possess communication skills, discussion may not be equally involving for all. Some

participants may feel threatened due to lack of experience in the subject of discussion or the technique. Thus, the dominant ones may lead the group leaving the less experienced behind.

2. If the participants do not have requisite knowledge and experience, discussion may not be useful.
3. It is quite time consuming. Trainer may not always have ample time in hand to use the method.
4. It may s not suitable for all kinds of objectives.

Application

This method can be used for planning, problem solving and decision-making.

Role of Trainer in Discussion

A trainer leaves the conventional role of an expert and becomes facilitator of learning. He has to motivate and help the learners in opening up and running the discussion process.

1. Trainer should create positive climate by arranging physical facilities like circular seating arrangement, stationary and aids for recording ideas. He should introduce the members, if they are not introduced already.
2. Trainer introduces the topic of discussion clearly so that the participants find it relevant to their knowledge and experience and feel motivated to pour their ideas.
3. Trainer must indicate the time limit and expectations from the participants in order to help them complete their task unaided.
4. Observation of the process must be done closely with appropriate tools and techniques to analyse the process. This requires specialized training to record all the relevant data regarding group behaviour and debrief them in a manner that participants can learn
5. Do not intervene in between unless it is very much essential.
6. Comment at the end about the points emerging out of the discussion as well as the process of discussion to deduce lessons.

Role of Leaders in Discussion

1. Keep the discussion on track by avoiding unwanted deviations.
2. Encourage the silent ones to speak
3. Motivate all to contribute.
4. Inspire members to give details wherever needed.
5. Encourage analysis of generalizations.
6. Summarize from time to time.
7. Resolve conflict.
8. Check monopolization of dicussions by the voluble ones.
9. Set norms so that everyone can contribute freely.
10. Test the opinion of all before concluding

Role of Members in Discussion

1. Each member should participate in discussion.
2. Speak one at a time.
3. Listen to the contribution of other members.
4. Do not negate the ideas of others.
5. Do not become egoistic or possessive about your ideas and force others to believe it.
6. Build group harmony by respecting right of others to speak

Points to be Observed Regarding Group Process

1. Initiative of members to participate and give new information/ideas.
2. Degree and quality of participation.
3. Influence exerted by individual members.
4. Communication skills of participants.
5. Co-operation and/conflict among members.
6. Norms formed by the members.
7. The way members of the group make decisions.

In order to provide comment to the participants both about the task being discussed and the process (of discussion) the trainer should record his observations on structured format with comments for precision in recording

Demonstration

Demonstration is a method of training in which the trainer shows step by step process of a method/operation and describes it.

Demonstration is very useful for teaching skills. Skill refers to the ability to do a task. Mastery in skill requires confidence in the technique and efficiency to perform task with investment of minimum time and energy. This can be achieved by showing the correct way of operation in such a manner that one can get confidence to try it out himself. This is where lecture or discussion cannot help.

Strong Points of the Method

1. Stimulates interest of learners because of audio-visual and action effect.
2. Builds confidence of the learners in the method by seeing the process.
3. It is very good for teaching skills.
4. Immediate feedback of demonstration is possible.

Weak Points

1. It takes quite some time and effort to carry it out as part of training process.
2. One may think it very easy by seeing the smooth handling by the trainer and gain false confidence, if opportunity to practice is not available.
3. It is difficult and time consuming to show a process and explain the steps together.

Application of the Method

Application of method is possible in cases where operation of a machine or using a tools to be taught. Trainers may be imparted skills in operating overhead or designing a transparency for projection by using demonstration method. Demonstration may be given to show how to use lecture or discussion method effectively in a training situation.

Tips for Using the Method

1. *Attractive Opening:* Start the demonstration with an attractive introduction to catch attention of the learners. Ask questions to learners regarding their experience with the method. If possible bring out the problems they face in using the method to be demonstrated. Keep the introduction brief but highlight the importance of the skill.
2. *Purpose Clarification*: Make the purpose of demonstration and its place in the session of training clear to the learners. They should clearly understand how the demonstration fits in the training session.
3. *Demonstration*: Keep the objects to be demonstrated well organized so that you can find when you need and time is not lost in arrangements. Organise seating in a way that they have no inconvenience in viewing and writing. Check equipments beforehand to avoid last moment embarrassment. Arrange the materials beforehand in sequence for easy handling. Begin the action part in a manner that you help the audience to follow each of your action. Tell learners what you are doing, why you are doing and then perform it. Look at the audience; point out the process so that everyone understands what is being done. As you complete a stage give extra information about thee materials, measurements, equipments and techniques to satisfy query of learners.

   Demonstrate the acts slowly. Be natural and enthusiastic. Repeat the important steps. Relate the process/step with the outcome so that learners can understand the cause and effect. Precautions or special care needed for conducting the practice must be told beforehand. Encourage learners to ask questions at the demonstration stage to clarify doubts. This will also provide feedback if everyone has understood.

   Good demonstrations must be supported by relevant audio-visual aids like charts, posters or models to make ideas clear.
4. *Initate Critical Discussion.*
5. Invite views of the learners about the practice being demonstrated.

Ask them about the easy and difficult steps. Problems perceived in doing the task should also be brought to the light. Application aspects of the demonstration should be discussed thread ware

6. *Encourage practice by learners*: Invite few learners to try. Motivate them to do the operation and correct soon to convey the appropriate way. Show special techniques of doing to emphasize on development of appropriate skills. Give enough time to practice.
7. *Close in a planned manner*: Do not leave demonstration abruptly after showing. Repeat key points. Emphasize on important techniques and application points. Show the finished product or achievement and create positive feeling. Convey any special warning or safety measures and application point for future use.

Role Play

Role Play is a method of training in which participants act out certain roles in an informal manner without script.

Strengths

1. It helps in bringing sensitivity among participants about certain behaviour.
2. It changes attitude of people.
3. It explores delicate human relations.

Weaknesses

1. It requires competent trainers to facilitate role-plays effectively.
2. Participants may take it casually and not pay serious attention.
3. The whole situation may get very artificial if not handled appropriately.

Method of Using Role Play

1. The trainers should prepare the participants before hand to play the role effectively.
2. Ask the selected participants to play roles.

3. Give signal for stopping the play.
4. Encourage the audience to share observations.
5. Analyze reactions of participants critically to lead to conclusion

Field Trip/Visits

Field trips or visits enable learners to see at first hand, or try for themselves, some of the things they have learnt in a training course.

To be effective and successful, field trips or visits need to be incorporated into the training programme with a clear sense of purpose. This in turn requires some effort on the part of the trainer:

- ☆ To communicate clearly the objectives of the activity from the start (or in advance),
- ☆ To plan and administer it effectively; and
- ☆ To provide ample opportunities for discussion and debate afterwards.

Not so infrequently, however, visits and field trips turn out to be mere 'social' events. The possibility of this inherent drawback, which faces even the more serious and well though-out outside activities, cannot be taken lightly even by the more competent trainers.

These activities may take any of the following two common forms:

Visits

Visits are mainly opportunities for learners to observe. Despite its passive nature, this nevertheless has the advantage of enabling an activity or project to be seen in terms of its environment. Some common, and useful, visit sites are:

- ☆ Meetings(village assemblies, user groups or associations gatherings)
- ☆ Exhibitions (civil works, handicrafts)
- ☆ Trade fairs and markets
- ☆ Project sites, etc.

Field Studies

Field activities can be turned into active learning experiences (involving 'doing' rather than merely 'seeing'). This can, for instance, be a valuable experience for participants in a Survey Methods course dealing with various data gathering techniques. Participants can be asked to prepare and test experimental questionnaires designed for a particular purpose (such as measuring rural household income or livestock ownership), or conduct interviews (with key informants) to gain insights into environmental problems of a given region. For Project Planning, Monitoring and Evaluation, too, there are many purposeful field study possibilities, which can enrich theoretical and abstract classroom discussions. For instance, if used in conjunction with the trainees' project work, the learning value of a field trip can be enhanced several fold.

Whether intended as mere observations, or designed to involve 'action', field trips are bound to require a certain amount of trainer preparation ahead of time–particularly if their educational value is not to give way to their entertainment appeal:

- ☆ Some written material may be needed for adequate guidance of visitors and project writers.
- ☆ The administrative arrangements may appear to be burdensome, yet prove to be worthwhile For instance, advance communication of details of participants, the course, purpose and the expectations of the visit, etc to the receiving institution can often make visits much more fruitful.
- ☆ For best results, participants need to be briefed beforehand and debriefed afterwards.

Advantages	*Limitations*
☆ Seeing is believing; doing is remembering	☆ The entertainment aspect may dominate, or even overwrite, its educational value
☆ Greater confidence in relating theory to practice	☆ time may be a real constraint (especially for short courses)
☆ Extends the learning process to environments that are difficult to capture in the classroom	☆ Elaborate arrangements may be needed ahead of time

Brainstorming

Brainstorming is a method used in small groups to generate creative ideas without inhibition. It is helpful to break the monetary and use resources/experiences of participants in solving a problem.Unlike group discussion and problem solving, emphasis is on encouraging every member to come out with ideas without criticism at the beginning. The idea is to benefit from diversity of experience, collective thinking and synthesis of contribution in an open and accepting atmosphere.

Strengths

1. It is quite interactive, as people get involved in the process
2. It is non-judgmental, as it does not allow for criticism or negative reactions from the participants
3. It is quite creative experience
4. It is often usefukl to solve problems through collective efforts.

Limitation

1. Group size–neither too small nor too gig
2. Integration of ideas generated requires high skills on the part of the trainer
3. Not all brainstorming exercises prove equally creative and innovative

Steps in Using Brainstorming

Prepare the Participants

Since the method assumes that all the participant can contribute significantly a feeling of importance has to be created amongst participants. Trainer should arrange comfortable seating facility preferably in circular fashion. He should introduce and welcome the members. Depending upon the understanding of trainees about the group, he should use informal conversation or exercise to make them relaxed.

Introduce the Session

The trainer can introduce the problem in a way that participants can relate with their experiences. He should clarify the rules. Every member has to contribute ideas without any hesitation. Negative criticism or judgment of ideas should not be expressed in the beginning. The idea is to generate as many ideas as possible. Even irrelevant and vaguely related ideas are encouraged. In the beginning the generation idea is more important than their relevance or logic. All the ideas should be recorded on flip chart or white/blackboard for all to see.I recording the ideas give importance to the words uttered by the members.

Rechecking the ideas

After every member has expressed views and no more is left, the trainer should go through the list of ideas to check if there is any repetition of ideas or vagueness in expression. All such ideas should be identified and explain the purpose is to make all the members understand and fell the ownership of the contribution made.

Analysis of Ideas

Now the group has to decide ways and means to make sense of the data recorded. So for if they are too many and divergent some criteria may be decided to classify them in groups. The merit and demerit of ech group can be debated critically to meet the objectives participants can be divided into smaller sub-groups to analyze different categories of solution.

Action Planning

This is the ultimate stage to bring consensus on the solutions sought. Taking into account the discussion made so far, concrete decisions are taken.

The method encourages creative ideas, free and fair discussion, rational analysis and evaluation. The method can be used to generate confidence among group member to generate workable ideas.

Chapter 17

Use of Teaching Materials and Aids

Audio-visual aids are instructional devices which help in stimulating the sensory organs as eyes, and ears and give the trainers, varied experience which enhances the quick comprehension of messages by them. It has several benefits over the traditional oral communication:

Creates Variety

Oral presentation becomes weary after certain duration. Therefore it is recommended to change stimulus after a gap so that variety could be created to maintain the trainees interest. Audio-visual aids create variety and retain public interest for long duration of time.

Visualisation

Trainees visualize the facts, process, skills and story through different audio visual aids. This visualization effect has significant advantage over the oral description of the same fact even in most interesting manner.

Makes Abstract Things Concrete

Projected or non-projected aids make demonstration of facts and figures possible providing a visual and concrete image of abstract things.

Retention and Comprehension much more

Even illiterate farmers retain and understand the abstract facts in more efficient way through the use of audio visual aids.

It is said, " *Words have wings and they vanish in thin air if you don't make the audience visualize the expressions behind it. What remains are the images and not the sound.*"

Variety of Audio-Visual Aids

A variety of teaching materials and aids are available like:

- ☆ *Class room aids:* Charts, posters, chalk-board, white board, bulletin board, flip book, models, maps.
- ☆ *Print media:* Folders, news letters, training manual/guide, hand-outs, worksheets/exercises.
- ☆ *Audio, video:* Recorded Cassettes, Film strips, transparencies, video film, television programmes.

Selection of Aids

A teacher should activate all senses." Wherever learning takes place the principle of teaching and learning remains the same *i.e.* all raise the question " how can communication be effective in transferring information and knowledge?". Generally when selected and used properly, teaching aids can help in the following ways:

- ☆ The interest of the audience can be maintained if the mode of presentation is varied. It could be difficult to concentrate for long on what someone is saying but referring to a wall chart or illustrating with slides, for example, can help to maintain the attention of audience.
- ☆ When information is presented in more than one sense (sight and touch, for instance), more is taken in, better understood and easily remembered.
- ☆ Difficult processes and concepts can be explained easily and clearly with the help of diagrams and charts. For example, an extension agent can teach a farmer about the life cycle of an insect using drawings or pictures of the insect.
- ☆ Pictures can have more immediate impact on emotions than words.

1. Select aids according to ability and interest of participants.
2. Select aids to match with learning activity.
3. Maintain balance in use of aids.
4. Select aids that complement each other.

The choice of materials call for considerations about various factors.

Teaching	Level of interaction desired
	Objectives
	Convenience of trainer
Student	Level
	Liking
	Group size
Message	Content
	Complexity
Practical consideration	
	Cost
	Electricity
	Equipment
	Space
	Manpower

Need of Audio-Visual Aids

These materials have different purposes

Teachers

Need materials to use for reference, communicate, support training session. *e.g.* Training guide, manual, visual aids, taped audio/video.

Students

Need to work on exercise-sheets, as reference material to take home and use directly in work situation as transparencies.

Media Use in Teaching

Principle of Media Selection

A teacher should activate all senses. Wherever learning takes place the principle of teaching and learning remains the same *i.e.* all raise the question " how can communication be effective in transferring information and knowledge?". Generally when selected and used properly, teaching aids can help in the following ways:

1. The interest of the audience can be maintained if the mode of presentation is varied. It could be difficult to concentrate for long on what someone is saying but referring to a wall chart or illustrating with slides, for example, can help to maintain the attention of audience.
2. When information is presented in more than one sense (sight and touch, for instance), more is taken in, better understood and easily remembered.
3. Difficult processes and concepts can be explained easily and clearly with the help of diagrams and charts. For example, an extension agent can teach a farmer about the life cycle of an insect using drawings or pictures of the insect.
4. Pictures can have a more immediate impact on farmers' emotions than words.

Chalkboard

One of the oldest aids to teaching, known since past four hundred years, chalk board is known for its ubiquitous presence in class rooms. It is a vehicle for variety of visual materials. Different name chalkboard, black board or green board knows it. There may be a combination of black, green and bulletin board for different purposes as well.

Purpose/Uses

1. Write important points to support oral presentation
2. Put illustrations for visualization of facts
3. Serve as base for recording students' inputs and help them in practice

4. Serve as surface for displaying non-projected visuals such as photograph, chart, graph, diagram, etc.

Problems in Use of blackboard

Black boards are so common in classrooms that they are often misused. Students face many problems due to poor board work by the teacher. Some of the common problems in use of black board are as given below:

1. Unclear writing or scribbles on the board which are not clearly visible.
2. Simultaneously writing and talking by the teacher.
3. Teachers often look at the blackboard rather than facing the students.
4. Teachers obstruct the view of the student by standing before the board
5. Teachers write a word or some points from somewhere or draw picture on the board which students can not connect with the lecture.
6. Irregular painting and lack of care of the board pose problems of poor visibility.
7. Glare on the board due to some source of light or inappropriate positioning of the board.
8. Writing on the board without erasing the materials not being discussed currently creates distraction

Points for Appropriate Use of the Board in Teaching/ Training

1. Plan board works in advance along with the text so that all important points can be put on the board for reference and discussion. Students get time to record and absorb key points if they are written briefly on the board. Unplanned writing results in hasty writing and erasing, which confuse students.
2. Start the class with a clean board. Erase previous writing or unwanted materials as you begin the class.
3. Write only key points only on the board.

4. Let the materials stay for some time to allow students to note and absorb the key points as they are being discussed.
5. Erase the materials not being discussed.
6. Cover the materials (facts or drawings) to be used later.
7. Face the learners and do not show your back by looking at the writing on the board for reference.
8. Stand at the side of the board and do not obstruct the view of the learners.
9. Use neat writing.1 to 1.5″ high and 1.25 inches thick letters are clearly visible to the back rows even in a 32′ long room. Separated letters and non-capitals are easier to read.
10. Use colored chalks for effect.
11. Check the visibility of the board by going at the back.
12. Use drawing aids like stencils, rulers, cutouts, etc. for illustrations.
13. While positioning the board look from several angles to see that there is no glare on the board.
14. Get the board regularly painted.The materials on the board should be erased with chalk board erased using upside down movements.
15. Now a day a combination of black and green boards as well as bulletin boards are put up in classrooms for making different uses.

White Board

Now a days classrooms are equipped with white boards which are also called as marker boards or multi-purpose boards. They can be used for more than one purpose. They are dustless and hygienic. They are preferred around computers as dusts can harm computers. They require special erasable markers which are non-permanent in nature. Felt eraser is required to erase the surface soon after use. Markers are available in eight different colours.

Purpose/Uses

1. It can be used for similar purposes as a black board in class/training room but has more clarity.

2. Use of a variety of colours markers can add to effectiveness of writing
3. It may be used as surface for projecting films, slides and overhead transparencies.
4. Some of these boards have steel backing and can be used as magnetic board for display.

Instruction for Proper Use

- ☆ Erase the writing as soon as the work is over. If marks are left for a longer period of time it will be difficult to erase and the white surface may be damaged.
- ☆ Special markers for whiteboard need special care. They should not be left open as they dry up quickly. They should be tightly closed soon after use. Even if the solvent base dries, it should be kept upside down and shaken vigorously to make it work.
- ☆ Wipe the board clean with a soft damp cloth. Soapy detergent can be used for further cleaning. The board should be rinsed with clean water and dried with towel.

Copy Board

It is high-tech version of the multipurpose board. It is a sort of electronic white board. It looks similar as white board but may be used for several purposes. It has multiple screen or frames which can be rolled back and forward.

- ☆ Teacher can write in advance and frames can be revealed at appropriate time.
- ☆ The device makes reduced size paper copies of what is written on board in ten second by pushing appropriate buttons.
- ☆ By copying the materials on paper, it is possible to erase the material and move ahead without losing valuable materials.

Bulletin Boards

This means a surface on which bulletins or news/information/announcements of general interest can be displayed. There are bulletin boards of different sizes with provision to hold pins, books exhibits and other materials.

Purpose/Use of Bulletin Board

1. It may serve as display surface to post materials to support instruction.
2. It may be used to decorate the room/office with photographs, pictures or tastefully picked up materials.
3. It may be used to motivate creativity of the learners to put up organized materials.

Desirable Features for Bulletin Boards

1. Writing should be legible from a distance
2. Lay-out should be attractive and balanced
3. The materials should be related with the objectives
4. A combination of facts, pictures drawing and other materials can add variety.
5. Decorative materials should be used to enhance attractiveness of the board like colored papers, tapes, pins

Cloth Board/Flannel Board

Cloth boards are stationary or mobile made up of cloth stretched over sturdy material such plywood or cardboard. Flannel cloth is often used because objects with rough backing like sand paper may stick on them.the cloth is usually of dark colours like blue, black or gray.

Purpose/Uses

- ✰ It is useful to present materials in sequence, one after another, to maintain curiosity
- ✰ Many different colours may be used.
- ✰ Symbols, pictures, letters may be used.
- ✰ It is impressive and dramatic.

Chart

Chart is a very popularly used graphic aid for communicating facts and illustrations to a group of people. It is used both in formal classroom instruction and training. It should contain only major ideas.

Advantages of Using Chart

1. It is easy to prepare. Teachers and trainers can make charts for personal use with little training
2. It is cheap to produce. The materials required cost little. No equipments are needed.
3. It supports teachers to introduce, present and summarize the lesson.
4. It complements use of other aids.
5. It helps teachers/trainers to structure their talk

Limitation

It has very short life span. It is difficult to store for a long time.

Types of Charts

Charts are of different types like organizational charts, flow charts, tree charts and tabular charts.

1. *Organizational charts* deal with structure or line of command of an organization.depicting hierarchy of personnel at different levels.
2. *Process or flow charts* deal with sequence, procedure or different steps in a process.
3. *Tabular charts* depict numerical data in tabular form for easy grasp of the relationship between two or more factors.
4. *Timeline charts* show chronological relationships between events over time.

Graphs

Graphs are often used to present information in easy and appealing manner not through words alone. Graphs are of different types like line graphs, bar graph, pie graph and pictorial graph.

- *Line graph* depicts relationship between two factors.
- *Bar graph* shows differences in quantity of one factor over another.

- *Pie graph* is drawn in the form of a circle by taking the circle as 100 percent. Different factors are shown as part of the circle by using different colours.
- A *pictorial graph* represents units through simple drawings to create interest.

Flip Chart

Flip chart consists of series of loose paper sheets stitched together at the top and which can be flipped one after another.

Advantages

1. It helps in sequential presentation.
2. It is cheap to produce.

Slide Projector

Slides are small individually framed photographic positives which are projected on screen. They are made from 35 mm film framed in 2x2" mount.

Advantages

- Used for both individual and group study.
- Can present vivid realistic illustration.
- Combined (synchronized) with audio-visual aids, it can provide audio-visual effect like video (if used with music, recorded voice and other sound).
- It helps to reduce the story with the aid of image, thereby help in illustrating major points carefully.
- It can be stopped at one point to discuss, and ask questions. Even one can go back and look at slides to sum up.
- It can be used again and again and can be rearranged.
- It is useful as self-instructional material.
- It is easy to produce slides from the film- using camera.
- They are relatively cheaper and can be sequenced alternately.

Features of Slide

- Attract and hold attention

- ☆ Add visual dimension to clarify points/show realistic
- ☆ Flexible in arrangement
- ☆ Self- instructional material

Practical Hints for Production and Use

- ☆ Make sure that the picture quality is satisfactory and the letters are readable in case of using a text.
- ☆ Select and sequence slides in a logical order to make the presentation clearer. You may use a lighted selection box to arrange.
- ☆ Make sure that you know how to operate the machine and ask for assistance in case you are uncertain.
- ☆ Identify objectives of your central presentation.
- ☆ Organize information in logical sequence.
- ☆ Visualize all the major ideas in the text and make a list of slides to show all the ideas.
- ☆ Plan visualization from general to specific. Use story board technique to plan frame by frame.
- ☆ To make story board draw out boxes for image, ideas and text to be said alongwith the slide.
- ☆ Choose one box for one slide frame/step.
- ☆ Draw image and ideas in form of simple sketches.
- ☆ Manage time factor so that audience can see each slide for 10-15 seconds and hear the narration. It will take about 10 minutes in presenting 50 slides.
- ☆ Review, modify and finalize the story board.
- ☆ Obtain a good camera and choose right film. Slides are made with special type of film (Slide film).
- ☆ Take 2-3 exposures of each composition to choose the best. Use colour meaningfully, limit number of words on slides. Use action scenes. Close-ups can be taken by keeping art work on copy stand and adjusting even light.
- ☆ Process film soon.
- ☆ Sort slide, get them framed and put in appropriate folder.
- ☆ Make list of slides and put appropriate caption.

Overhead Projector

It runs with the help of overhead transparencies which are transparent sheets of papers on which information can be written and presented on screen. These transparencies are made of cellophane, acetate, glass and infrared film.

Strong Points

- ☆ Projection can be easily interrupted.
- ☆ Continuous eye contact with the audience.
- ☆ Different ways to produce and use sheets.
- ☆ Sheets can be presented in a chosen order.
- ☆ They can be used in normal light.
- ☆ They help in showing facts and illustration.
- ☆ Well-planned transparencies make good impression.
- ☆ Can be used again and again.
- ☆ Can be cleaned and re-used.
- ☆ Can work as black board for spontaneous writing.

Weaknesses

- ☆ Danger of presenting too much information in a short time.
- ☆ Lecturers get easily addicted to the medium and tend to use it almost regularly and indiscriminately.
- ☆ If the user is not cautious, it is easily vulnerable to damage.
- ☆ Movement can not be shown.
- ☆ Projection equipments and source of power are must.
- ☆ Needs darkened. Can't be used in open.

Practical Hints for Using it

- ☆ Prepare sheets carefully by using key-words and short sentences.
- ☆ Write clearly but not too much.
- ☆ Use the right colors (black, dark blue, deep red and deep green).
- ☆ Switch off the projector when you don't use it.

Using Transparencies

- ☆ Preview materials and practise presentation.
- ☆ Use a pointer to direct attention to the section of the transparency you are referring to.
- ☆ Cover the transparency with paper to show only the part being discussed. This will make them curious about next step and avoid unnecessary distraction.
- ☆ Use felt-tip pens or special OHP markers to write.
- ☆ Show each transparency for appropriate time.
- ☆ Explain rather than read what is written on transparency.
- ☆ Encourage question and answer.
- ☆ Give copies of transparencies on paper to audience for reference.
- ☆ Neatly typed and photocopied transparencies with illustration make presentations professional.
- ☆ Store them in file or envelop by keeping a blank paper cover with each transparency.

Motion Pictures

Films refers to the celluloid material on which series of transparent image, when projected 24 frames per second, is perceived as moving. Illusion of vision is created by persistence of vision. Films of different width are available which are used for distinct purposes.

☆ 8 mm	Instructional
☆ 16 mm Film	Instructional
☆ 35 mm	Movie theatre
☆ 70 mm	Cinemascope

Advantages

- ☆ Uses sight and sound.
- ☆ Motion can be presented.
- ☆ Enlarge/reduce object.
- ☆ Past can be reconstructed.
- ☆ Easy to reproduce record of event to be used again and again.

- Influence attitude.
- Abstract relationships can be understood through colour, motion and sound used in motion picture.
- Provide enjoyment.
- Editing of reality makes presentation more attractive.
- Invisible objective can be shown.

Photograph

The word photograph is derived from two words 'photo' means light and 'graph' means to draw. This simply means to draw with light. Photograph is nothing but a combination of light and shadow. It involves a camera, film, light, a subject and a photographer. But the biggest component in this process is an idea of what to photograph. It has been said, that *Photographs are created in minds and not in camera.* The trainers need photograph for distinct purposes. Therefore, a mental exercise for needed image is to be done before one goes for actual shooting. Getting photographs according to our distinct needs is not an easy and simple job.

Advantages

- Its easy to capture, document and carry photographs at any on-campus and off-campus teaching and training sites.
- Showing photographs does not need any machine, power supply and other support services.
- Its repeated and prolonged use is possible.
- Being cost-effective, a set of photograph may be changed and enhanced at any interval without much financial considerations.
- It is ready to use medium in which editing is not needed.

Recommendations for Effective Photograph

- Develop well-structured lesion plans for each session. Decide the interventions, through photograph according to the lesion plan and prepare a list of photographs needed.
- Work over each proposed photograph to decide the specific requirement and content.

- ☆ Take purposive action photographs depicting required details.
- ☆ Try to go as close as possible to the object. Avoid uncounted details.
- ☆ Objective and purpose and faces.
- ☆ Always care about the light source and backgrounds. Keep simple background.
- ☆ Keep people busy while having the photograph.
- ☆ Use your flash gun for in-door photography.
- ☆ Choose right film for photographs. Remember, that you have to prepare big enlargements.

Video

Video is a visual medium that includes one additional dimension to traditional audio-visual aids *i.e.*, motion. Thus, video becomes nearer to peoples experiences and perception. It provides a support to skill building and behaviour modeling. Thus, its an effective medium of training for enhancement of skills.

The video format has covered an unbelievable journey form VHS mode to digital mode. Now, it is easier to handle a digital handy cam with internal memory. Data transferring to computer, editing through user-friendly software, voice transfer and incorporation of animated captions in between the video has also become easier. Use of video camera in skill-based sessions and its instant replay to evaluate the performance is also possible now through is also possible now through support of LCD projector, computer or TV monitor.

Recommendation for Effective Video Films

1. Video proves to be an important tool to be used in teaching as it leads to significant change in skill and understanding of the students.
2. Video can be used as an effective tool for self analysis in applied arts, teaching and communications.
   - According to the learning objective, instructional design and strategy, the role of video has to be decided.
   - The video format (VHS, Super VHS, Betacam, digital) depends on the purpose as well as budget. For in-

house, one-time use in labs for demonstration and documentation etc, its better to restrict to less- costly formats as VHS, Super-VHS etc. But simple VHS recording can't be effectively edited with music and voice overlay and it is not suitable for professional productions.

- Prepare a script of your story through developing and building the concept.
- Treatment of concept is the next step in which a narrative description is developed which tells how the programme will appear on the screen. It includes the chronological sequence of audio and video, locations, shots, music, graphics, sound, actors and other related decisions.
- A rough draft of all the shots and the supported audio component is sequenced in story board. It is complete visual depiction of all the shots that link to form the story.
- This story board and final script lead to shooting script that becomes a operational guideline for cameraman, artists, director and other related parties.
- Ultimately, the shooting is done and the recorded material is edited on linear/non-linear editing units.

Audio-recordings

Skillful combination of speech, sound, silence and music in trainings have their significant implications. Its simplicity and cost-effectiveness makes this media very prospective to be used in teaching and trainings. Recorded message supported with music and dramatization not only enhances the comprehension but also break the monotony in training. More over, it may be used in outdoor conditions, in broad-day light and at such locations where power supply is not available. This is their reason behind expansion and effective role of broadcasting for agricultural and rural development.

Innovative and vivid use of sound recording is possible to enhance quality of trainings. Recorded cassettes and line recordings could be used in trainings as an effective medium of enhancement of knowledge and a understanding.

E-learning as a Component of Teaching

E-learning is a term, being used very extensively in present context, along with other terms as online-learning open and distance learning, web-based learning etc. We often use these words interchangeably. E-learning in simplest and crude way, is internet based learning. Internet has appeared as the extraordinary tool of information for sharing world wide knowledge. E-learning provides opportunity to access this wealth of information at a time, place and pace of learner's convenience. E-conferencing and on-line courses are new dimensions to this e-learning process. One additional dimension of e-learning is flexibility to use the information in media format of user's choice. With this conceptual basis, the e-libraries have come-up through-out the world having millions of e-books and references available at their ends in easily accessible formats.

E-learning as a Virtual Support to Teaching

All developed universities throughout the world, take support of e-learning as a well-defined capsule of their teaching package. It works as supportive and follow-up mechanism of each session where we learn skills and explore all related details and similar experiences on net in the end of the session to enhance understanding in world perspective.

Students are expected to come prepared on certain topics through E-learning before they come to the class for discussion over the given assignment in present system of education. It enhances the interest and interaction of the students and provides a holistic view of the content. It's all the more important in the context that the world is changing fast and to keep-up the records according to its pace, e-learning is an effective tool to provide world scenario at the doorsteps of the user.

Multimedia for Effective Teaching

Multimedia is a combination of text, pictures, sound, video and animations to present a coherent programme. Obviously, it is more result-oriented and effective than any one media used at a time. It can be directly utilized for practicing and mastery over skills, problem solving, providing simulation, individualized and cooperative learning and providing life-like situations to learners to stimulate them to learn and crystallize their problems as well as solutions.

The biggest advantage of multimedia as a tool of teaching is, it utilizes natural information-processing abilities, already lying in brain. Human brain processes visuals and movements in easier way than text materials. Even such texts, which have few good illustrations, are processed in better way. Multimedia provides a series of life-like situations including sound, video, graphics and animation. Interactive multimedia provides the liberty of using information according to the user's choice. For example, trainees of Apiculture may get full illustrations of bee boxes, types of bees, their sounds, activity in the box alongwith data and figures related with their economics as well as outlets, packed in one multimedia presentation.

Thus, multimedia has the strength to simplify even the most complex and crude content into simple, understandable and interesting format which is totally user -friendly.

In Nutshell

Programmed Instruction

It is particularly useful as in enrichment activity. It can help provide highly motivated students with additional learning experiences. Programmed materials have been proven to be very effective in remedial instruction. It can be an effective means of providing classwide competencies in skills. It is used for mass teaching.

Still Pictures

These are effective in providing knowledge to a group. But the level of interaction of this media with the learner is low.

Motion Pictures

These are useful in teaching skill and for changing attitude of the people. It is appropriate n mass teaching. There is medium level of interaction between the media and the users.

Television

It is mostly used for providing knowledge to a large no of people at a time with medium level of interaction.

Audio-Recordings

These changes the attitude and adds further knowledge with

the present level of the group. The interaction with the participants is of medium level.

3-D Objects

Such objects has medium level of interaction but they enhances the knowledge of the target people. There are mostly used at mass level.

Printed Textbooks

They have got low interaction with the learners. These are good at providing knowledge to large no of people.

Chalk Board

This is very useful in providing teaching to a group and also widens their horizon of knowledge. The level of interaction is low.

Poster

It is helpful in changing attitude of the masses but has low interaction with them.

Charts

It also helps in changing attitude of the people but has low interaction with them.

Dramas

They are used in teaching at mass level with medium level of interaction. They are useful providing knowledge.

Radio

This has got immense importance in providing knowledge to a large audience though it has got medium level of interaction.

Chapter 18

Games and Exercises for Enhanced Learning

Public dealing for maximization of learning is challenging. It aims to help people acquire new abilities or modify old ones. It is not easy. People do not like to change. They resist efforts to change as they may not be conscious of the need or sticking to old habits is easy. Learning also means that they (learners) are ignorant. Moreover, perception of traditional classroom or odd memories of child hood learning make learners uncomfortable. Learning is lifelong and continuous. We keep learning all the time intentionally or unintentionally. People associate learning with classroom learning. But learning in life is entirely different for adults. Adults are mature and responsible people. They are critically aware of their talents and responsibilities and want to seek opportunities for their expression. They learn through actual experiences and interactions with their immediate environment. Adults want to solve their immediate problems and enhance their performance. The degree of maturity and related attributes may vary among people. However, trainers should understand the characteristics of adults that distinguish them from children or young learners and deal differently with them. This also forms the basis for the newly emerging field of andragogy or teaching methodology for adults. Andragogy is based on five basic assumptions about characteristics of adult learners, as described below:

Adults have Self-Concept

Adults have images of self as capable and independent being. They are responsible and capable of making decisions for themselves. They can plan, work and achieve objects of their desire. Thus, they like to choose their own directions. Too much of instructions, orders or dictations are not liked by them. Thus, it is clear that adults do not need undue interference in handling their life and work. They need respect, acceptance and motivating physical environment to get over the initial hesitation. Initially, they are anxious about the learning environment. Unpleasant memories of learning situation and experiences in childhood make them silent and anxious. They are afraid of ridicule and humiliation. They can open only if there is no punishment, ridicule or other forms of humiliation. Trainers should understand this characteristic and arrange stimulating physical and psychological climate to open them up. It is better to keep instruction non-directive.

Adults have Experience

Adults bring with them unique experiences of life by virtue of their background, resources and roles. They have learnt to deal with problems and gained experience. Experience has become integral part of their personalities. They are rightly proud of it and associate it with their identity. Ignoring their experiences means ignoring them personally. Trainers who do not take into account the experiences of learners make them feel rejected and out of place. Thus, trainers must plan to include experiences of learners as valuable resources to be exploited fully for making learning multi-faceted and lively. Experiences are the basis to relate new learning to the past. There is yet another side of this. People are also conditioned by experiences gained in the past. They tend to work/behave in a certain way and resist any change as they feel comfortable in the way they have been doing things.

Thus, it is imperative for trainers to adapt the training to the realities of past experiences of learners and provide many opportunities of mutual sharing and critical discussion. Attempts must be made to help learners understand the influence of conditioning. Orientation sessions should be conducted to create conducive environment to relax and get relieved of anxieties and past beliefs and assumptions. New experiences should be provided and participants should be encouraged to discern meaning for themselves.

Adults have Readiness to Learn

Every period of growth in human life creates an opportunity to learn. Adults are also driven to learn because of their roles and responsibilities in personal, social or work lives. When a new responsibility is to be carried out, they are motivated to learn and perform better. They are ready to acquire attitudes, knowledge and skills needed to carry out their responsibilities effectively. It is thus necessary that trainers recognize the teachable moments when adults are eager to learn. Curriculum should be focused on real life concerns. This will draw adults towards the subject and help put efforts to achieve goals. Learning environment will be quite conducive if people with similar needs are grouped together. Understanding people's real needs and priorities are first and foremost requirements before initiating training.

Adults have Orientation to Learning

Children come to schools to learn things and they use the acquired knowledge later in life if opportunities come. There is no hurry or pressure to use. The orientation is futuristic. However, adults are keen to use new learning in their immediate context. It is a question of life here in the present tense. They want to meet their immediate needs. Thus emphasis is on application. There is no postponement of the decision to act. This calls for practical orientations to training. Trainers should approach their curriculum and design of training with problem-orientation of learners. There should be ample scope to discuss problems of application of knowledge and skills being taught.

Adults can Learn Differently

Adults are different as learners and deserve differential treatment. There is a general feeling that adults cannot learn. It is quite common to hear from elderly people that they can not learn now. It is true that physical capabilities diminish with growing age. Eyesight starts getting dim with age. Many people become hard of hearing and memory also gets weaker with advancing age. However, adults maintain intellectual capacity to learn. Memorization may be difficult but they can learn all the way through reasoning and comprehend things. Those who keep intellectually active can learn all through.

Experiential Learning through Games and Activities

Experiential learning is an innovative approach to make people open for new learning.We know that class room learning is dominated by lectures and learners remain passive on-lookers in the process. In real life adults learn through actual experiences. They observe self in action, think, feel, analyze and draw lessons for life. Experience becomes the basis for generating learning. Experiential learning also means the same thing. It is an approach to integrate natural learning process in a teaching or training session. It is centered on learners. Experiences are organized to make learners active and involved. We have to act as facilitator to provide experiences relevant to them and help in analyzing the experience in order to gain useful insights. Learner is active all the time. We merely enable their learning by providing opportunities of involvement, practice, observation and reflection.

Simulations or games are imitations of real life situation. They provide opportunities to make decisions based on life-like situations. We process the behaviour of the participants after the game to derive lessons out of the game. Games are not merely the means of entertainment. They model real life situations and enable the participants to deal with them and make decisions. Understanding the motive behind them and processing the behavior of participants in order to derive lessons is the main part of the story and this is where the role of a facilitator comes into being. A moderator or facilitator is the one who facilitates the entire process. His main role is to give a concrete shape to things, linking the motive to real life situations and drawing an outcome.

Introduction to Games

Tell me and I will forget, show me and I may remember, involve me and I will understand.

The best way to learn is by doing where the learner can actually get involved, experience things and understand the underlying concepts in a better way.

As per the *Cognitive Experiential Self Theory* (CEST), by Professor Seymour Epstein at the University of Massachusetts, we have an experiential mind and a rational mind. Our experiential mind learns directly, thinks quickly, pays attention to the outcome, and forgets slowly. Our rational mind learns indirectly, thinks deliberately, pays attention to the process, and forgets rapidly.

Games link directly to the experiential mind by making the participants face life like situations. The Debriefing section, accompanied with the Games helps in linking those situations to the actual theme, and formation of abstract concepts. Thus overall it makes the process of learning more effective as well as dynamic.

Why Games ?

The shell must break before the bird can fly

Learning Made Fun

If the mind is free and lighter, it is more open to learning, allowing the easy permeation of new concepts and new ideas. Games help in creating such a fun-filled ambience where the participants find themselves at ease thus enabling them to get free from all fears, anxieties and other hindrances. They can be their true selves – more adaptive as well as responsive, open to learning and open to growth. Thus the process of learning becomes much easier.

Not Abstract, but a Clear Picture

Apart from this, games give a practical dimension to all abstract ideas and thoughts. The participants can actually experience the different situations and conditions which otherwise would not have been clear had they just heard about them, or seen a slide show related to them. By the means of games, they are made to undergo various circumstances where their manner of dealing the situation, decision making capabilities- everything talks a lot about their personalities and revels their inherent creative potentials. During the whole process, the participants also gain good deal of learning experience which they can apply in similar real life situations in future.

Unlearning and Relearning

Learning is all about unlearning and relearning:

Breaking of pre conceived notions and making space for the entry of new. This is what games are good at. They also make the learning process inside out. The learner is not forced to learn, he learns on his own, through the experiences he undergoes.

Experiential Learning and the Role of Games in it

Experience is not what happens to a man, but what a man does with what happens to him.

Experiential learning as the name suggests is learning through experiences. In contrast to conventional learning techniques where the key role is of the trainer, experiential learning methods give the reins of the process into the hands of the participants. They are made to undergo various experiences where they can actually perceive things in a different light and act accordingly. The facilitator acts as an observer, deeply observing each and every step of the participants. His role becomes effective at a later stage during the debriefing process.

David Kolb in his book *The Importance of Learning Style* gave the **Experiential Learning Cycle** in order to describe the process of experiential learning.

As per this cycle, experiential learning happens through four stages:

1. *Concrete experience*–Games, simulation games and exercises provide this learning experience.
2. *Observation and reflection*–The process is observed critically. Individual participants are made to reflect over the process to understand things behind the crust. Churning and reflection is done.
3. *Forming abstract concepts*–A set of data linked to actual learning evolves through the churning process and the experiences of participants.
4. *Testing in new situations*–applying those in new situations in their peer group, office home. Applying those concepts in varied situations.

David Colb's Experiential Learning Cycle

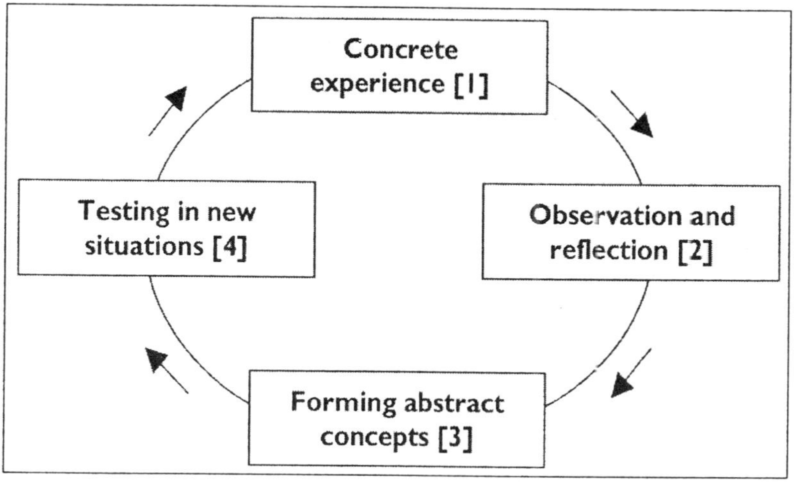

Games integrate all these four points together and are the best means for experiential learning. The mode is necessarily participative *i.e.* each individual has to indulge in the ongoing activity where they are assigned specific task to accomplish. The team working, decision making, analyzing and strategizing skills of participants lead the process and help in accomplishing the desired goal. The whole process provides a good learning experience to the participants. The debriefing section then leads to a better understanding of things where the facilitator makes the participants share their experiences and links the reflections and feedbacks to the main purpose and theme, forming abstract concepts which can further be tested in different such situations.

Debriefing

Debriefing forms the major part of the Games. It is this segment which gives the reins back in the hands of the facilitator where he instigates the participants to reflect and give their feedbacks over the learning experiences obtained from participation in the games. He then links all the experiences accumulated during the process to the abstract concepts of the main theme, giving a proper direction to the flow of events. The game or activity section of the Game accounts for only the first part of the experiential learning cycle *i.e.* 'Concrete

Experience'. It is the Debriefing section that completes the experiential learning cycle by facilitating 'reflection' from participants that lead to the 'formation of abstract concepts' which can further be 'tested in various situations'.

What to Observe

The sensitivity of the facilitator plays an important role in debriefing process. Who wins and who loses in the game is overtly visible. The facilitator has to see something that is not directly visible. Understanding human psychology, involvement, attachment and detachment of human tendencies

Not to confine on content but to go for the process.

Overall observation is about going beyond the scene, to understand cause rather than effect.

Reflection

Making the participants reflect, provoking them to reveal about the activity and the event. Gradually driving towards the process-why and how is it important? Who did what? Making them reflect over their feelings and emotions. Leads to linking the process with personal lives and experiences.

Out of all the observation and reflection, the facilitator must be capable of deriving abstract concepts. They must not be superimposed. The participants must feel them as their own.

What the facilitator observes must be shared with participants in between their reflections. It must be done step by step in order to make reflections comparatively more strong and efficient.

Role of a Facilitator

Facilitator is one who moulds the process, giving it the required shape.

- ☆ The main role of facilitator is empowering people, not merely extending any knowledge or information.
- ☆ He has a restricted role, not impressing audience, goal clarity. Never test and evaluate.
- ☆ Create congenial environment, enhance comfort level of audience and participants.
- ☆ Making the snails come out of their shells.

- ☆ Lead the process to accumulative learning.
- ☆ Approach is never one sided or exploitative, rather its cumulative and empowering.
- ☆ Involving all the participant. Keeping an eye over every individual and trying to stretch themselves to the comfort level of participants.
- ☆ Have indepth sensitivity to understand language of silence rather than language of words.
- ☆ Approach is like handling a scalpel, rather than holding a butcher's knife.
- ☆ Equal hold over content and process.
- ☆ Not like sage on stage or a saint to chant, but like a peer to cheer.
- ☆ Doesn't control. Lose control.

Types of Games

Games may be used in different environments and for different purposes. Personality development workshops, classroom teachings, trainings – all may utilize games to make the learning process more interactive as well as effective. Development of team skills, initiative taking and decision making abilities, group dynamics, communication skills etc. can also be done quite efficiently and flexibly by the means of games.

Games can be categorized as:

1. Indoor games
2. Outdoor games

Icebreakers

Icebreakers are mostly used for the initiation of sessions, for letting the participants get acquainted with each other, shed their inhibitions, open up and adjust themselves to the environment. In contrast to the other games, they do not involve a debriefing segment.

Example

Everyone is given a slip of paper when they come in that has a famous person, character, or something that could be "grouped" on it. The goal is for people to find other people that are a member of the same prearranged "group" and get together. Then once people find everyone in their group, they can do something as a group like discussing over a topic, a picture, a quotation etc. In the end, one representative from each group is asked to come forward and represent his/her group.

Team Building Games

Team building games help in understanding the working of a team, group dynamics, role of each individual, the strengths and weaknesses of each member that support or hinder the process and areas of improvement.

Example

Group is divided into teams each having 4-5 members.

Members of each team have to imagine themselves as different personalities like doctor, engineer, student, soldier, old man etc. traveling on a ship which wrecks during the voyage.

Only one of them can be rescued.

Now in this situation team members have to decide among themselves that who is the one to remain alive and why?

Debriefing

Learning to make decisions together is the main objective of this activity. In this process of decision making, few of the things that matter are:

- ☆ *Each one has his unique role*. Different people in the team play different roles–some participating actively, some just mute spectators, some dominating the discussions and not giving opportunities to others, some taking the form of a lateral leader–guiding and encouraging others etc. but it is the combined effort of all that team reaches to a particular decision. A decision is successful only when each member expresses his views freely and supports as well as challenges the ideas of others in order to produce the best outcome.
- ☆ *The role of a leader*. One of the members emerges out as the lateral leader – one who guides the overall process, gives direction to the ensuing discussions, resolves any conflicts arising, finalizing the discussions to reach to a common

conclusion approved by all. He has good people skills, is considerate about everyone and is committed to team approach.

☆ *Creativity of the team members.* Members present different views and ideas regarding who must be saved and why. While some think according to the normal patterns, some are those who bring out very vivid, innovative, out of box ideas supporting their cause.

Creativity Games

These games bring out the inherent creativity of participants. Such tasks or activities are assigned to them that require out of box thinking. Participants are made to think different from the normal patterns.

Example

One of the participants is asked to come in the front and draw anything on the board. Rest are called one by one and asked to add to the drawing anything they wish. But each person can add just one thing to the drawing.

Debriefing

Two things that emerge out through this process are:

1. The role of an initiator
2. Out of box thinking of the participants.

In this game, it is generally seen that whatever the first person thinks and draws, the other ones go according to it. Thus initiator plays a big role in every event or activity- be it a group discussion, or meeting etc. He is the one who gives direction to the activity and can create "track" for others. Hence the final outcome depends heavily on his efforts and his attitude.

Another thing that matters here is the innovativeness or out of box thinking of the other participants. Some will be those who'll follow the same track as set by the initiator – just adding something supporting the previous one. But few will be those, who'll try to create their own ways–adding something that will be very new, different from the previous ones. And these are the ones who introduce freshness and add a different dimension to the otherwise monotonous patterns.

Trust Games

These games aim at the development of mutual trust between team members.

Example

A person has to fall freely from a height of 3-4 feet. To catch him, the team members create two parallel lines with people in them facing each other. Person going to fall stands on the elevated platform with his back towards the rest of the team and falls backwards.

Debriefing

Experiencing trust is the main motive of this activity. It allows the group to experience trust as the faller and to provide support and create a trusting environment. The way a person falls shows how much he trusts his fellow team mates. For example, if one is fearful or hesitant while falling, it shows his less trust for fellow team mates.

As participants catch teammates, they develop team support, trust and caring which creates a sense of team bonding.

This activity lends itself to discussions around the following topics: letting go, trusting teammates, providing support, experiencing support, clear communication.

Precautions

Few things must be kept in mind while using the games in sessions, classes or workshops:

- ☆ All Games can't be used in all situations.They are location specific as well as audience specific. Some are meant for outdoors and some just for indoors. Some have no criteria of age group, but some are meant only for a particular age-group.
- ☆ Debriefing plays a very important role in games. It should be brief but in-depth and must not deviate from the main theme.
- ☆ Role of facilitator is very important in the process. Everyone can not handle every type of situation. If wit and humor are required during the session or debriefing, then a faciltator who is having it in nature can only be able to handle the session efficiently.

- ☆ Goal clarity is must for the facilitator. He should be very clear about the purpose and the motive behind the game. Only then will he be able to lead the process to the desired destination.
- ☆ The ultimate outcome of game should be the empowering of participants. Each and every participant must feel fully charged, energetic and confident at the end.
- ☆ Accessories and tools to be used for the game have to be pre arranged so that events flow smoothly and no chaos situation is created after the initiation of session.
- ☆ Any new game, before its use in the session, must be practiced earlier. It should be practiced fully *i.e.* upto debriefing and not merely the activity.
- ☆ Every observer has limited span of control *i.e.* he is able to handle and observe only few participants. Thus adequate number of observers should be used to observe each participant.
- ☆ The facilitator / observer must use a small diary to record the observations. It is better to use video recording for better analysis and reflections.

Chapter 19

Role Play in Method of Teaching

"I hear I forget; I see I remember; I do and I understand"

Confuscious

Role play may be defined as the way one behaves in a given position and situation. Role playing as a teaching methodology is the conscious acting out of discussion of the role in a group. In a classroom, problem situation is briefly acted out so that the student can identify with the characters. In other words, role play is a form of dramatization in which a student spontaneously performs what he/she feels, thinks and imagine, about a particular role under specific conditions. Gone are the days when teaching methods used to be passive; listen and read. What about doing? Today, it is fashionable to "do". Good teachers "do" to promote an atmosphere of active learning among the students. They want the students to realize that learning is an interactive process which can be fun as well as academically rewarding.

"One of the most well established rules of of social science is the "participation hypothesis" : As participation grows, "ownership" grows and that bond opens the participant to new ideas, innovation and learning........ The more active are the students in determining and absorbing their own learning, the more they learn" (Mann 1996).

Concepts of Role Playing

- ☆ Role play must be considered as an experimental procedure, a method of learning to perform roles more adequately. It is an act, or a spontaneous play.
- ☆ Role play is sometimes referred to as socio – drama. It is an unrehearsed dramatization of a problem in which the members of a class without scripts extemporaneously portray how they would react in a given situation.
- ☆ Role play when used in a classroom refers to a spontaneous drama in which students assume certain roles and play them as they see them.
- ☆ Role play is an unrehearsed dramatization in which players attempt to make a social situation clear.
- ☆ It is spontaneous enactment or dramatization of a situation in which students impersonate people of various ages and occupations. The action comes from the students' creative use of their feelings and imaginations.

A careful review of the definitions stated above reveal that the chief characteristic of Role play is spontaneity of presentation. Besides it pertains to assume behaviour of particular persons act of being someone else. The main difference between role play and socio – drama is that role play seeks to provide comprehension or understanding in a particular role for the person playing that role whereas socio drama on the other hand seeks to utilize role play as a means of devising solution to a prescribed problem situation.

Purposes of Role Play

1. To examine a delicate problem in human relation.
2. To explore possible solution to an emotionally laden problem.
3. Provide insight into attitudes differing sharply from those of the participants.

The ideal size of class to present role play is of 25 viewers; however, it can be demonstrated to a larger group. The limitation with large group is that the reaction might not be feasible or may not be meaningful.

Steps in Role Playing

As a teaching technique, role playing is based on the philosophy that *meanings are in people,* not in words or symbols. If that philosophy is accurate, we must first of all *share* the meanings, then *clarify* our understandings of each other's meanings, and finally, if necessary, *change* our meanings. This implies changing the concept. The self concept is best changed through direct involvement in a realistic and life related problem situation rather than through hearing about such situations from others. Creating a teaching situation which can lead to the change of self concepts requires a *distinct organizational pattern.* One helpful structure for role playing follows more than the simple acting out of the roles. Though it has a little formal structure, it is made up of a series of steps as per the following:

Selecting the situation

Choosing the participants...... Role taking

Setting the stage

Acting out the situation

Evaluating the presentation

Re-enacting.....Role reversal (from step 1)

Selecting the Situation

This needs meaningful and comprehensible situation. Give description of the background or contextual setting of the situation. A brief description of the roles played can also be given to have clear mind and understanding for the significant interaction later on. The teacher must *identify the situation clearly* so that both the characters and the audience understand the problem at hand. In casting the characters, the wise teacher will try to accept volunteers rather than assign roles. Students must realize that acting ability is not at stake here but rather the spontaneous discharge of how one thinks the character of his role would react in the defined situation.

Choosing the Participants

They can be suggested by teacher and student. Students can themselves volunteer for the roles. Teacher needs to guide the selection to avoid students taking roles that will make them feel self conscious or uncomfortable. Players may be instructed publicly so that the audience knows what to expect or privately so that the audience can interpret the meaning of their behavior. Be sure to allow for creativity of the actors within their character roles and do not over structure the situation.

Setting the Stage

Give few minutes to discuss the roles among themselves. Let them get the feel of situations and personalities. Allow for furniture setting. Quick 'buzz' groups can help refine decisions and add new points to presentation. Teacher should give concrete guidelines to the audience to observe what educationally significant and valuable learning is.

Acting Out the Situation

Allow participants to follow their own course of spontaneous action. Allow feeling of freedom for creating a realistic atmosphere. In case of misinterpretation of role, teacher should intervene. Teacher needs to impress student that they are playing the roles, are empathizing and not portraying their own feelings and attitudes. Interaction requires the pupil to make necessary adjustments to the entire situation in process in terms of the requirements of the role. This will help to gain better understanding of the roles and relationships as well as better awareness of what they are doing.

Evaluating the Presentation

As soon as the presentation is over, the role players and observers are given an opportunity to discuss and summarize learning gained. Role play is evaluated in terms of fullness of situation, meaningfulness, strong and weak aspects of situations and aspect of role, the ways in which conflicts are resolved. Besides the role players, auxiliary coaching from teachers and audience is necessary. The audience is just as much involved in the learning situation as the actors are. In the analysis and discussion time, the audience should provide possible solutions to the realistic problem situations which surface.

It is important to *evaluate role playing* in the light of the prescribed goals. Categorizing behavior is often overdone and gets in the way of the learning process. Evaluation should proceed on group and personal levels, raising questions concerning the validity of the original purpose. Throughout the entire process it will be necessary to deal with certain problems which arise in role playing situations. The backward, silent member must be encouraged to contribute. Create an atmosphere in which he is unafraid to share ideas, confident that no one will laugh at his contributions or harshly criticize his conclusions.

The overbearing monopolize must be curtailed in the discussion phase of role playing lest he dominate the group and thereby quash the dynamic, Solving this problem may require some personal counseling outside of class. Tension and conflict in the group may not always be bad. Sometimes these elements act as a stimulant to thinking. There is such a thing as "creative tension," and it is frequently found in a role playing situation as group dynamic emerges.

Re-enactment

Following the discussion of various roles and situations, other scenes might be enacted to show different set of actors as suggested by actors themselves, teacher or audience. A willingness to repeat and improve a performance is important interaction for the effectiveness of this technique. At the end of the discussion, the group should collectively measure its effectiveness in reaching towards the solutions to the role problem posed at the beginning. The techniques of role playing afford another approach to involving students in their own learning process toward the clarification of self concepts, evaluation of behavior and aligning of that behavior with reality.. You can see why this is a desirable approach to classroom procedure for the teacher. Successfully used under the guidance of the teacher, role playing can be an effective instrument in classroom teaching.

Values of Role Play

- ☆ Role playing allows people to make mistakes in a no threatening environment. They can test several solutions to very realistic problems, and the application is immediate. It also fulfills some of the very basic principles of the teaching-learning process such as learner involvement and intrinsic motivation. A positive climate often results in which one can see himself as others see him.

- ☆ The involvement of the role playing participants can create both an emotional and intellectual attachment to the subject matter at hand. If a skillful teacher has accurately matched the problem situation to the needs of his group, the solving of realistic life problems can be expected.
- ☆ Role playing can often create a sense of community within the class. Although at first it may seem a threatening method, once the class learns to share a mutual confidence and commitment to the learning process, the sharing of analysis over the role situations will develop camaraderie never possible in conventional teaching methods such as the lecture.
- ☆ Promotes awareness and impression of the thought, attitudes and perspective of some one else thus promoting appreciation of other person's views.
- ☆ Role play provides for enjoyable realistic learning experience in order to solve the problem creatively.
- ☆ Provides opportunity to express and release emotions without feeling self conscious and without making the problem appear to be his/her.
- ☆ Increases communication skills, presents emotional subject matter in an impersonal manner.
- ☆ Role play develops understanding of human relations and group behaviour.
- ☆ Provides experience in acquiring empathy thus develops inter – personal competence.
- ☆ Helps students to gain insight into their own behaviour and develop more adequate patterns of behaviour
- ☆ Role playing helps in changing attitudes toward the members of the groups.
- ☆ Role playing allows for undesirable and desirable social behaviour without making the situation appear too instructive.
- ☆ This method of teaching enables the students, especially the role players to identify with real people.

- ☆ Role playing can be used with students of most ages. The complexity of the role situations must be minimized in using the method with children. But if we keep it simple for their limited attention spans, role playing can be used even in teaching preschoolers.

Problems in Role Playing

- ☆ Perhaps the major drawback to teaching by role playing is the insecurity of class members. Some may react negatively to participating in a situation which will be discussed and possibly criticized by other members of the class. And role playing takes time. The class discussion of a five-to-ten-minute role playing situation may extend to several times the length of the situation itself. Sometimes extremely beneficial results may accrue. At other times, because of ineffective performance on the part of the players, or mishandling on the part of an unprepared teacher, the outcome may only be a superficial rehash of what everyone already knows about the problem.
- ☆ The relationship of the people in the group is a crucial factor in the success of role playing. At times it may emerge as a negative factor. For example, previous interpersonal difficulties experienced by group members may arise in class to corrupt the role playing situation. Also, if the group has people of different status, they may be reluctant to become involved for fear of being humiliated before the members of the class who are smarter or more popular.
- ☆ Actors may become self conscious, non – spontaneous and non – responsive if they cannot project themselves in real situation.
- ☆ Calls for great deal of time for setting up situation, describe character, act the scene, discuss feelings and depict actions.
- ☆ It is unsuitable to display physical skill as it not only aims at showing feelings and emotions.
- ☆ Students may live their parts so vividly and successfully that there is a danger of dissipation of the personality.
- ☆ If a role player enters into the situation to become subjectively a part of it, it is difficult to revert to an objective examination of what it all means.

Conclusion

The educational benefits to be gained from using role – play are immense. With careful planning in construction, organization and running of the role – play and with added bonus of a new technology, there is no reason why many of the weaknesses of this method might not be overcome.

Today technology has progressed so much that the traditional form of role play has been mixed with the new technology. The web – based role play might be most effectively engaged in when used in conjunction with rather than a replacement of the more traditional face- to- face approach of this traditional teaching method.

Chapter 20

Teaching Packages for Improving Quality of Teaching

Excellence in teaching cannot come by merely wishing. It calls for systematic approach to instruction. Casual teaching without advance preparation, thinking or support materials ends in nothing. No one is benefited. There are wide variations in manner of teaching. Same course may be taught differently by two teachers. It is indeed alarming when nature, level and amount of contents also differ. This puts a question mark on teaching standard. Differences in experience of teachers also create differences in outcome of teachers. Thus, there is a need to think of a system through which such differences can be minimized to a great extent. One of the perpetual complaints of the students is that the contents of same course by two instructors are so different. Another common problem is leak of prepaventurs of new teachers because of lack of standard materials. Teachers teach for two to three decades but all their experience of teaching go away with them. The academic departments do not have any document regarding teaching approach unless some books are authored. Thus, there is need for such documentation to make teaching systematic, constant and standardized. Development of teaching package provides opportunities to overcome these problems.

What is Teaching Package?

A teaching package is set of teaching materials on CD, floppy or print that helps in effective learning from instructor to students. Teaching packages are also helpful to run the course smoothly. This includes course outline, miniature and digital version of visual aids, reading materials schedule of exams, exercises and handouts needed to run the course.

Advantages

- It provides standard materials to be used during the course. Teaching can consult and get ready to teach. One does not need to search all over again.
- Course is so structured that teacher has more time for interaction
- Ensure consistency in teaching due to adequate materials for each section.
- It is easier to hand over course to new instructor.
- Presence of teaching package will make it easier for new instruction to teach.
- It makes learn teaching easier.
- Teaching packages may be continuously upgraded.
- Handouts and overhead transpierces can easily be multiplied any readily used.
- Teaching performance can be effectively assessed.
- Examples of questions and elaboration of evaluation criteria may prepare students to face the examinations.

What Does a Teaching Package Contain?

A teaching package contains all the materials in hand and soft copy (on compact disc) that a teacher needs to teach a course effectively as below:

- **Course-outline:** Detail programme of each lecture practical along with suitable references may be mentioned.
- **Lecture notes:** Well-organized notes of each lecture may be includes.
- **Handouts:** Short key points with little elaboration may be prepared.

- **Visual aids:** Overheads, poster, slides and charts may be Xeroxed and kept.
- **Assignments:** Reading, writing or practical assignments and relevant details should be provided.
- **Evaluation criteria:** Distribution of marks and criteria of assessment for anwser script alone with sample papers can be provided.
- **Schedule:** Of exam, field visit etc. with instruction.
- **Reading materials:** Addition related reading materials can be Xeroxed and added.

In fact this is a step-by-step guide for a new teacher to conduct the class systematically. The documentation of teaching approach may help in preparation of teacher guide for specific subjects over a period of time.

Experiences of Teaching Package

Alemay University, Ethiopia with help and support from Larenstein Professional university, Netherland had started preparation of teaching packages for two departments *i.e.* Department of Agricultural Extension and Department of Animal Science. About 20 teaching packages were prepared by each department and put in print and CD form. Two copies of the packages were put in departmental libraries. A teacher can borrow the package for a semester and copy the materials. It makes the task of multiplying handouts very easy. However, teaching packages could not be evaluated and up dated requality. Besides all teaching packages were not of good value. Some of them looked like poor version of text books. Lecture notes were generally were not included in the package. Local examples and thorough explanations of materials were missing.

How Can Teaching Packages be Made More Effective?

On the basis of Ethiopean experience it can be said that teaching packages can be made more useful in standardizing teaching and providing support to new teachers. Motivation to teachers there is need to select teachers who have experience and interest in preparing teaching package. Willing teachers can do wonders provided they are given some incentives. In the Ethiopian cas, every teacher was given US dollars 900 per course to prepare teaching package.

Evaluation of Teaching Package

There is need to form an internal and/or external committee to critically review the teaching package in the light of the learning objectives and performance goals. All teachers may not be expert in educational technology. Thus, there is need for budget/project for evaluation of teaching package.

Infrastructure Support

Teaching packages need to be stored in departmental documentation center/library reading room where computer should be available. Student may be allowed to view the CD/folder/file in the reading room. Facility for copying may also be provided. Teachers need liberal access to computer Xerox facility, scanning and other reprographic facilities to copy materials from other sources for inclusion in reading materials.

Section 4
Good Teaching Practices: Issues and Experiences

Chapter 21

Application of Total Quality Management in Home Science Teaching

There is a growing and welcome realization amongst the developing countries that education is the key to development. The quality of manpower in any country ultimately determines the sustainable well–being of its people. Creation of social opportunities for all sections of society is a reflection of the progress of that society and education is the principal instrument for developing human capabilities. Education, as a liberating force, fosters growth, social equality and technological progress. Globalization and the emergence of a new society are more dependent on knowledge and information technology and have further underlined the importance of education in pursuing developmental goals. In education, good teaching is defined as an instruction that leads to effective learning, which in turn means thorough and lasting acquisition of the knowledge, skills, and values the instructor or the institution, has set out to impart. A college or university seeks to improve the quality of teaching by what it uses, does and delivers. The ultimate goal is to enhance classroom effectiveness in order to improve student learning more effectively.

As Home science is one of the subjects which aims at developing its students into wholesome personalities equipped with capacities and competences for fulfilling their niche in national economy, professional world, entrepreneurial arena, community development and in family life, therefore quality of education is of prime importance in this field. To maintain quality in Home science education we have to improve teaching because quality requires in identifying problems with existing academic practices and then applying a combination of sound educational and psychological principles to devise a better approach. Our concern in this paper is specifically with value of Quality Management in Home Science education, how Quality Management can be applied in education (the teaching and learning process) to improve the most crucial element of the learning environment.

Value of Quality Management in Improving Classroom Instruction

Quality Management is a collective strategy that has meaning only if it is agreed upon and implemented by the staff of an organization. Applying Quality Management terms to instruction in a single course by a single teacher may provide a good experience for the students, but it is not Total Quality Management. It should be a collective effort from an institute involving the faculty as well as the students. Thus it becomes necessary to identify emerging areas of Home Science and course programme should be formulated based on those areas. New ideas and innovations should be supported and implemented. Scientific and technological application of course programmes should be emphasized for the benefit of the society. The students should be exposed to problem solving aspect of education using their analytical, critical and creative thinking.

Strategies for Improving Teaching Quality in an Individual Class

It is not difficult to find semantic links between teaching and quality management. In the sections that follow, we describe several strategies known to be particularly effective to improve the instructional program at a university.

Writing Instructional Objectives

Instructional objectives are statements of specific observable actions that students should be able to perform if they have mastered the content and skills the instructor has attempted to teach. Well-formulated instructional objectives can help instructors prepare lecture and assignment schedules and facilitate construction of in-class activities, out-of-class assignments, and tests. Perhaps the greatest benefit comes when the objectives cover all of the content and skills the instructor wishes to teach and they are handed out as study guides prior to examinations.

Using Active Learning in Class

Most students cannot stay focused throughout a lecture. After about 10 to 15 minutes their attention begins to drift, first for brief moments and then for longer intervals, and by the end of the lecture they are taking in very little and retaining less. Attention can be maintained throughout a class session by periodically giving them something to do. Many different activities can serve this purpose of which the most common is the small-group exercise. Active learning exercises may address a variety of objectives. Some examples are as follows which can be followed according to the need of subject.

Recalling Prior Material

The students may be given one minute to list as many points as they can recall about the previous lecture or about a specific topic covered in an assigned reading.

Responding to Questions

Any questions an instructor would normally ask in class can be directed to groups. In most classes–especially large ones–very few students are willing to volunteer answers to questions, even if they know the answers. When the questions are directed to small groups, most students will attempt to come up with answers and the instructor will get as many responses as he or she wants.

Problem Solving

A large problem can always be broken into a series of steps, such as paraphrasing the problem statement, sketching a schematic or flow chart, predicting a solution, writing the relevant equations, solving them or outlining a solution procedure, and checking or interpreting the solution. When working through a problem in class, the instructor may complete some steps and ask the student groups to attempt others.

Explaining Written Material

It is a powerful activity for helping students understand a body of material. The students are put in pairs and given a text passage. The instructor calls on one or two pairs to summarize their explanations. A member of each pair explains each statement and the explainer's partner asks for clarification if anything is unclear, giving hints if necessary. The instructor may also help students if they lack somewhere.

Analytical, Critical, and Creative Thinking

The students may be asked to list assumptions, problems, errors, or ethical dilemmas in a case study or design find the logical flaw in an argument; predict the outcome of an experiment or explain an observed outcome in terms of course concepts; or choose from among alternative answers/designs/models/strategies and justify the choice made.

Generating Questions and Summarizing

The students may be given a minute to come up with two good questions about the preceding lecture segment or to summarize the major points in the lecture just concluded.

Using Cooperative Learning

Cooperative learning (CL) is instruction that involves students working in teams to accomplish an assigned task or assignment and produce a final product (*e.g.* a problem solution, critical analysis, presentation of information, laboratory report, or process or product design). Relative to students taught conventionally, cooperatively-taught students tend to exhibit better grades on common tests, greater persistence through graduation, better analytical, creative, and critical thinking skills, deeper understanding of learned material, greater intrinsic motivation to learn and achieve, better relationships with peers, more positive attitudes toward subject areas, lower levels of anxiety and stress, and higher self-esteem. In cooperative learning the instructor has to define the students the five conditions of cooperative learning. The results of cooperative learning include the following elements:

Positive Interdependence

Team members are obliged to rely on one another to achieve the goal. If any team members fail to do their part, everyone on the team suffers consequences.

Individual Accountability

All team members are held accountable both for doing their share of the work and for understanding everything in the final product (not just the parts for which they were primarily responsible).

Face-to-Face Promotive Interaction

Although some of the group work may be done individually, some must be done interactively, with team members providing mutual feedback and guidance, challenging one another, and working toward consensus.

Appropriate Use of Teamwork Skills

Students are encouraged and helped to develop and exercise leadership, communication, conflict management, and decision-making skills.

Regular Self-Assessment of Team Functioning

Team members set goals, periodically assess how well they are working together, and identify changes they will make to function more effectively in the future.

Formal cooperative learning is not trivial to implement, and instructors who simply put students to work in teams without addressing the five defining conditions of cooperative learning could be doing more harm than good.

Guidelines for Cooperative Learning

The following guidelines suggest ways to realize the benefits and avoid the pitfalls of cooperative learning.

- ☆ *Proceed gradually when using cooperative learning for the first time.*
- ☆ *Form teams of 3-4 students for out-of-class assignments.* Teams of two may not generate a sufficient variety of ideas and approaches, teams of five or more are likely to leave at least one student out of the group process.
- ☆ *Instructor-formed teams generally work better than self-selected teams.* Classroom research studies show that the most effective groups tend to be heterogeneous in ability and homogeneous in interests, with common blocks of time when they can meet outside class.

- ☆ *Give more challenging assignments to teams than to individuals.* The level of challenge should not be raised by simply making the assignments longer, but by including more problems that call upon higher level thinking skills.
- ☆ *Help students learn how to work effectively in teams.* One technique is to collect anonymous comments about group work, describe one or two common problems in class and have the students brainstorm possible solutions and select the best ones.
- ☆ *Take measures to provide positive interdependence.* Methods include assigning different roles to group members (*e.g.* coordinator, checker, recorder, and group process monitor), rotating the roles periodically or for each assignment.
- ☆ *Impose individual accountability in as many ways as possible.* The most common method is to give individual tests. It includes calling individual randomly to present and explain team results; having each team member rate everyone's contribution and combining the results with the team grade to determine individual assignment grades.
- ☆ *Require teams to assess their performance regularly.* At least two or three times during the semester, teams should be asked to respond to questions like "How well are we meeting our goals and expectations? "What are we doing well?" "What needs improvement?" and "What (if anything) will we do differently next time?"

Assessing and Evaluating Teaching Quality

Most institutions use only end-of-course student surveys to evaluate teaching quality. While student opinions are important and should be including in any assessment plan, meaningful evaluation of teaching must rely primarily on assessment of learning outcomes.

Conclusion

Agreeing on educational goals is only the first step toward formulating an academic mission. However, in the modern university, teaching has just become one of several important functions. This attitude needs to be changed as teaching should be the most important objective of an academic institute because it is directly linked up with the future of our youth. There is a crying need to improve the teaching quality of universities to enforce better academic discipline. If effective teaching strategies are known, these may simply be incorporated into class room instruction.

The measure of quality in education of a university includes the imparting of certain knowledge, skills, and (perhaps) values, a meaningful assessment process. Some efforts had been made in applying Quality Management in education, although most efforts are still in their infancy. Quality Management requires that it only works if it is total. The same requirement applies in teaching and learning that all members of the institution are committed to improve it. If a university wishes to move its organization toward a culture that embraces positive change, the university must lead the way. We must not only "talk the talk", but should also "walk the talk". So Let us "walk the talk" toward the new millennium.

Chapter 22

Veterinary Education in India: Issues and Strategies

Training was never an inexpensive preposition and when we talk about any professional training, such as training future veterinarians, we are referring to a serious expense of resources. The resources refer not only to the heavy burden on the exchequer but also of other valuable resources, including the man-hours, both of the trainers and trainees.

In the era of ever increasing global competitiveness the general expectation of the masses from any kind of service provider, has increased. Educational organizations being no exception to such expectation, have realized the need to compete on both national and international levels. This increased competition has resulted in wider range of choices for the potential students and shifted power from the supplier (the educational institutes) to the customers (the potential students).besides this, a huge amount of financial and other resources are incurred on making a professional-say a veterinarian.'Quality' in such professional education has become a critical competitive issue in this environment.

Indian Livestock sector plays a crucial role in our rural economy and livelihood. India has a large animal population and to fulfill the requirements of their feeding, breeding, management, health and hygiene, processing of livestock products such as milk, meat, large force of trained personals are required, besides this, there is a requirement of teachers for Human resource development and research in frontier areas of animal and veterinary sciences. Veterinary education started long back in India and with growing need of the society, education in all discipline of animal sciences spread in almost all Indian states and currently 37 veterinary colleges are functioning in the country, besides this, some ICAR institutes where veterinary education in various related disciplines of animal and veterinary sciences are provided up to Master's as well as Doctoral level. These students are being absorbed in academic institutions, extension department and management and industrial sectors and are performing very well. However, globalization and liberalization has brought new challenges for the country our education system including that of veterinary sciences needs major reforms in order to make out students most competitive in the era of liberalization. They need to be trained to take up the challenges sportingly and for this purpose we have to reform our curriculum up to post-graduate level so that students come to the expectations of the stake holders.

Veterinary Council of India and Veterinary Education

It would be worthwhile to mention about the conceptual approach of standardization and thus checking quality of veterinary education in India. Veterinary council of India is statutory body constituted under Indian Veterinary council Act, 1984 (52 of 19480. it is charged with the responsibility, apart from regulation of Veterinary practice and maintenance of Indian Veterinary Practitioner's register, to maintain minimum standard of veterinary education in the country. The Parliament while assenting on the bill endorses strongly that it would be mandatory for the council to regulate veterinary education as it form that foundation for regulating standard of veterinary practice.

In confident with this specific provision of the VCI Act 1984, minimum standard of veterinary education degree course (B.V.Sc. and A.H.) Regulations 1993 were laid down. VCI has definitely given a standard base for undergraduate course but several issue still remains to be addressed with special reference to the Post Graduate courses.

Current issues in Veterinary Sciences

Emerging diseases, foreign animal disease, agro-terror, and international food safety-all point to the importance of globalization of our profession. A recent large market survey Association (AVMA) stressed the importance of educating veterinarians in the field of international veterinary medicine. The fields demanding increasing attention. These are public health, epidemiology, wildlife and regulatory issues. Recent crises in international public health, *e.g.* Bovine spongiform Encephalopathy (BSE) in United Kingdom, avian influenza in south East Asia and Central Asia, Hendra and Nipah viruses in SE Asia, or Senere Acute respiratory Syndrome (SARS) in China, demonstrate how the line between human and animal disease is becoming increasingly blurred. It is obvious that specialist in comparative medicine, *i.e.*, veterinary are needed to meet the challenge of emerging international zoometric diseases. The curricula for post graduate programmes in animal health need to address these issues. There are many wildlife issues related to international agriculture and trade that need to be addressed by veterinary specialists. It has to be clearly understood that in the global ecosystem, understanding the cause and effect of one ailment and finding solution in isolation will be just a grace period for a greater disaster to strike. The development world is greatly concerned with the incidence of foot and mouth disease in Europe. On the other hand, the developing world of ours, are trying to understand and stay away from the threat of bird flu. Given the volume of people in the world today and the incredible interconnectedness of international trade in food, issues of foreign animal diseases, agricultural economics, and emerging diseases can only continue to grow. If we are not preparing tomorrow's veterinaries to deal with these problems, we are courting disaster for our profession. Veterinary colleges are beginning to recognize the tremendous need for educating students in such international matters. However, there are roadblocks to clear and hurdles to cross.

Reforming Post-graduate Curriculum

There is urgent need for advancing veterinary education. The mere revision of the course will not be sufficient. It has to include development of suitable modules, delivery system, role of different organs of education system, refresher programmes, skill enhancement programmes, scope of distance education etc.

Following board areas are worth consideration:

Updating of Curriculum

Keeping the curriculum contemporary as well futuristic is indeed a great challenges. It is a vital area in preparing the sciences and Technology manpower for tomorrow. Unfortunately, syllabus for most of the post-graduate programmes is too orthodox. Curriculum change is a slow process particularly when faculty members are feeling increasingly stretched. One effort, which appears plausible in our education system too is formation of clusters between the Universities. In this effort several veterinary colleges/ state Agriculture Universities group themselves together to form consortia for competitive federal funding. Each college is responsible for the development of several computer modules involving a specific disease or mechanism of control. The final result will be a complete stand-alone course that could be used by every school of veterinary medicine. At the beginning some of the important disciplines for post-graduate programmes can be selected for a trial.

Faculty Development

Members are responsible for designing, Implementing, maintaining and refining a challenging, contemporary curriculum that encompasses a variety of basic and clinical science courses, all designed to prepare students for challenging future for tomorrow, to bring a high level of challenging future for tomorrow. To bring a high level of enthusiasm, competence and commitment to their teaching duties, at the same time aspiring to serve as mentors, coaches and role models for students enrolled in the post graduate veterinary science curriculum. Faculty members are responsible for the curriculum content they offer the methods of student assessment they employ, and are responsible for determining whether students have demonstrated sufficient proficiency or mastery in each aspect of the curriculum. It is them who will have to instill that confidence in the students to face the challenges of the tomorrow's world.

Leadership Role of Dean (Academic Affairs)

A few associate Dean for Academic affairs should be nominated to become member of some related faculties charged with a number of administrative responsibilities, including general oversight of the curriculum and ensuring fair and impartial application of the academic standards in all matters relating to students academic performance. The Associate Dean, may, in the course of counseling and advising, consult with members of the faculty, administrators and other as necessary. He/she may also review and refer to a student's veterinary academic record and, when necessary, pre-veterinary academic records in an effort to comprehend the issues underlying a student's academic deficiency in order to help devise a suitable plan for remediation, when called upon to do so.

Inter- and Intra-Institute Clustering and Resource Pooling

in the present set up there are some institute and universities with experienced teachers, well-furnished laboratories and well-equipped libraries. At the same time there are also teaching institutions without such facilities to the desired extent, such institutions may be encouraged to from clusters among themselves in order to share their human resources and infrastructure. Such cluster of institutions in the same place/near by place can offer different optional courses so that the students would have a wider choice for electives, choosing a course from any of the institutions in the cluster Mechanism to evolve convenient timetable transfer of credits from one college to the other etc., will have to be evolved.

Collaborative Learning

This method of learning, in which small groups of students help each other to learn, will allow newer areas to be covered effectively till a suitable curriculum evolves ins these sear. Collaborative learning has been described as a process by where each member contributes personal experience, information, perspective, insight, skill and attitude with the intent of improving learning accomplishment of others.

Curriculum Design Workshop from Members of Board of Studies

Introduction of innovative teaching programs as well as up gradating of curricula and syllabi periodically is the prime function of the Boards of studies. Since the quality of teaching programmes is directly proportional to the professional conduct of these boards of studies, which must be facilitated through workshops on curriculum design incorporate such inputs as need-assessment teaching methodologies etc.

Distance Learning

Suitability of method of teaching to entire post graduate degree programmes in veterinary science needs attention. Certain course *e.g.* refresher courses, short term and certificate courses, Skill enhancement courses, short term and module wise certificate courses can possibly be more acceptable than those requiring more laboratory exposures. Obviously distant learning only through printed course material is obsolete. Virtual class room and virtual laboratory modules will have also to develop in an interactive mode.

Emerging Areas to be Included in Curriculum

The areas that are emerging fast in changing global scenario and which need to be suitably introduced in post-graduate curriculum of veterinary science specializing in Animal health disciplines are:

- ☆ Molecular biology, genomics and proteomics
- ☆ Genetic engineering related to diagnosis and vaccine production
- ☆ Modernization of vaccine production technology (GMP, GLP, DNA vaccines and Edible vaccines)
- ☆ Biological vector control
- ☆ Alternate traditional medicine
- ☆ Bio-engineering and bio-informatics
- ☆ Biosafety, World Trade Organization (WTO)and Intellectual property Rights (IPR)
- ☆ International trade including non-tariff barriers
- ☆ Management
- ☆ Certification and validation of products and policy for international acceptance
- ☆ Semen Banks of Elite Animals
- ☆ Risk assessment, risk monitoring and management of diseases
- ☆ Impact and economic of diseases management

- ☆ Animal ethics and animal rights
- ☆ Superspeciality hospitals

Post-graduate Curriculum in Clinical Subjects

In comparison of animal health management, hard core clinical subjects, *e.g.* surgery, gynae and obsterics therapeutic medicine, are fast losing out as favorites for post graduate studies.

We have not yet even started super specialization, *e.g.* ophthalmology or dentistry. There is an urgent need for attention to these areas. Even within the existing structure, incorporation/ updating of specialized courses in PG curriculum, *e.g.*, companion animal medicine and surgery, equine medicine and surgery, stress and pain management in animals zoological, exotic and wild life medicine, urban anima health management, emergency care and neurology, should not be too difficult. In a world where information is doubling every few years, reforms of curriculum is not mere up-gradation of curriculum. The road-map to quality assurance in veterinary education programme can possibly be prepared in combination of some of the points discussed above. The institutions and the faculty will also have to make clear distinctions between lecture and teaching. Collaborative participation of the private sector, corporate sector and foundation/trusts evolution of course or programme curriculum, diversification of resources, adequate financial support for R and D in learning centers are some other facets which in combination of the above can really prepare our students for tomorrow's world.

Chapter 23

Quality Teaching of Agronomy: A Reflection

"Good Teaching Inspires Flowering of the Mind"

Agronomy means, in general, agriculture, so it is the most important branch of the agricultural that deals with soil and crop management for higher crop productivity. Since the subject agronomy is related to management aspects, it requires special emphasis on practical aspects of soil and crop management for better understanding of students. The prerequisite of quality teaching is based on the synergy of teacher-student-teaching environment continuum. If any part of this continuum lacks, the quality teaching is affected adversely. Therefore, a good teacher should be highly qualified, experienced, good orator and most importantly enthusiastic to deliver the knowledge. Similarly the clients or students should be curious enough to imbibe the knowledge. The teaching environment also play important role in the teaching process, therefore the conducive teaching facilities should be available for better exchange of knowledge.

Agronomy deals with mainly principles and practices of crop production, weed management, irrigation water management, rain fed agriculture, crop ecology and also economic aspects of crop production. Besides, the infestation of pest and diseases and their proper management can not be ignored during successful crop production. Therefore, the subject Agronomy is very wide and joins hands with different disciplines of science, management and agriculture that make Agronomy one of the most important and useful subject of agriculture but on the other hand, there is a general feeling of our graduate students that all the courses of Agronomy are lengthy, data based, boring, mind boggling and requires much more practical based knowledge for better understanding of the subject. It is always difficult task to make any subject easy and interesting particularly in the eyes of the students because the present generation of our students needs ready made study material with least efforts and it makes students lack of subject knowledge, irregular and also irresponsible. Therefore, the issues related to 'student-teacher-subject matter-teaching environment' have been debatable and will remain as such and their best possible solutions might have emerged in future too. With the view point of the present scenario of our higher education system and its related pros and cons, the undercoated points may be beneficial not only for quality teaching of Agronomy but also to make it interesting.

- ☆ Vibrant and dynamic course curriculum
- ☆ Practical orientation
- ☆ Modern infrastructure
- ☆ Need based research
- ☆ Interactive teaching
- ☆ Assignment presentation
- ☆ Motivation and confidence
- ☆ Employment generation
- ☆ Internationalization of the subject
- ☆ Extra curricular activities
- ☆ Teachers' Training.

Vibrant and Dynamic Course Curriculum

The course curriculum of any institute has its own impact and significance that reflects the dynamism think tank of the institute. The course curriculum of the institute should be very vibrant and dynamic so that the recent development of science and arts may be included in the syllabus timely. Agronomy is a whole agriculture, so the large syllabus has been divided in to many specialized courses and the syllabus of each course has been updated. Further the courses need to be modified time to time and make it up to date with the recent development in the subject. The curiosity and enthusiasm should be created among the students about the subject by metaphorically speaking with soft heart because the students are full of interest and wonder and also remember that knowledge has to be imbibed, not imposed. The hearts are like flowers, they remain open to the softly falling dew but close in the violent downpour of rain.

Practical Orientation

The practical orientation is the most important requirement of the quality teaching of the subject agronomy because to make student acquaintance with different types of crops, weeds, insects and diseases, agricultural implements, seeds and laboratory equipments need either well equipped laboratory, live crop museum, herbarium files of crop and weeds separately and functional equipments or students should be allowed to visit some other important teaching and research institutes having related facilities for better understanding of different climates, their crops, weeds, pest and diseases and also site specific crop and soil management problems and their specific solution. It is very much required that students should be well exposed to field conditions and problems related to crop production. Hence, the agronomy needs its course curriculum modified providing 50 -50 % weight age to both theory and practical for making the subject interesting and quality teaching. For better understanding and enriching the subject knowledge, it is desirable that students may be attached either with some progressive farmers or with university or government farms for at least one crop season. It will provide good opportunity to our students to involve with daily agricultural activities, crop planning and budgeting, scheduling irrigation, weed management, nutritional deficiencies, pest and disease infestation and their scientific management, harvesting and post harvest techniques, and also marketing of the produce. Besides, students can realize the problems of our farmers and farming systems. This type of scientifically organized training will make our students full of subject knowledge and develops confidence among students for being excellent farm manager in future.

Modern Infrastructure

Presently the modern and hi tech infrastructure of the educational institutes including well furnished AC's class rooms, good hostel facilities, OHP, LCD, computers, internet facilities, library facilities, well equipped laboratories and well established research farms with all important farm implements etc. are the source of satisfaction of both parents and their wards and attract to good students. The Hi-tech. class room facilities help not only in delivering effective lectures but also improve the grasping power of the students. Therefore the modern facilities should be easily accessible to our students as well as teachers for effective teaching of the subject. Most often it happens that the high tech. instruments or implements are available in the institutes but remain decorated piece in the fancy rooms or almirahs. This trend has to be changes for the benefits of the students, teachers and also the institutes. The outburst of information technology has opened the doors for E-Learning, E-Teaching and Education etc that should preferably be encouraged.

Need Based Research

The current trend of research work carried out by our M. Sc. and Ph.D. students for their degree programs is very monotonous and sometimes boring, hence it needs to be tilted towards need based and site specific research on project mode. The students should be encouraged to conduct their research work on farmers' fields rather than on university research farms. This will provide a wonderful opportunity to our graduates to interact with farmers for different kinds of problems like farm and household as well as socio-economic problems. In addition, students may feel happy and proud for their research work that may solve many important farm problems easily. The recent trend of professionalism calls for excellence, high grade performance and also clarity in grasping the problems and its solution. Therefore, the experiential learning on project mode should be taken care of for higher education of agronomy for composing our graduates with high grade professionalisms.

Interactive Teaching

Interaction of students with subject experts, teachers and students of other institutes, farmers and agri-business personals should be encouraged through group discussion, seminars, symposia, field visits and educational tours for exchange of quality knowledge of the subject. The interactive teaching will create the wonderful atmosphere particularly to students and I hope the students will open their hearts and inherent potential of the students will come out under interactive teaching.

Assignment Presentation

Sometimes the class teaching is not sufficient in PG teaching. The students should be motivates for more learning by providing them home work, assignments, class notes, cited references of journals and books, news capsules in advance and then discussed in the class. Further, the allotting of assignments related to the subject topics and its presentation in the departments is very useful technique of effective teaching. The writing and presentation of assignments help students in many ways like:

- ☆ Force study of the subject
- ☆ Increase subject knowledge
- ☆ Acquaintance with new development of the subject
- ☆ Expose with new books and journals of the subject
- ☆ Opening of hearts and hesitation
- ☆ Increase interaction
- ☆ Increase thinking.

Motivation and Confidence

The motivation towards excellence in subject matter and confidence in the Agronomy are two important points that must be inculcated among the students because both help in shaping their gesture, modesty, sincerity, honesty, devotion and overall charming personality that are essential components of good professionalisms. The high confidence alone solves 50 % problems. Therefore, the orientation lectures of the subject should be arranged for students emphasizing the importance of the subject, future scope and related employment opportunities, so that students may be motivated and opt Agronomy preferably rather than unwillingly. The orientation lectures of great personalities and knowledge of the subject will create motivation and confidence building among students.

Employment Generation

Inclusion of interesting and employment oriented courses/ topics like remote sensing and geographical information systems

(RS and GIS), bioinformatics, biotechnology, crop diversification, traditional wisdom, allelopathy (Residual study) etc. may certainly help to attract good graduates for higher studies in agronomy. Besides, these topics are not only researchable burning issues presently but also are highly potential in generating employment opportunities to our agronomy graduates.

Internationalization of the Subject

In our present scenario of higher education, the specialization starts after graduation. Our good graduate chaps prefer either job or management courses in search of better job opportunities.

opportunities and high salary structure. It is thought that only poor stuff get admission in PG and Ph.D. programs because no other options left for. If you think about the international agriculture, Agronomy is the main branch that includes Agronomy; Plant Breeding and Soil Science and provides better job opportunities that makes big difference compared to our system. Besides, the specialization begins from graduation degree itself and continues up to the highest degree in China and some other countries and prepare their graduates with full of subject knowledge and confidence. Further, their graduates have to work in the same field either as a teacher, manager, researcher and they really perform wonderfully. Why not we do adopt the internationally accepted course curriculum in our universities for making our graduates ready and well equipped for international market?

Extra Curricular Activities

The extra curricular activities may provide extra food and energy to the puzzled minds and make them emotionally bound very fresh and refreshing. Therefore, the extra curricular activities should be integral part of 'Teaching and Learning Process'. The activities like 'Competitive Quiz', 'Games', ' Debate', 'Cultural Shows', 'Picnics', 'Fun Fares' etc. should be intermingled in between the process of teaching and learning. These activities work like a tonic to the minds. Besides, it decrease the gaps between students and teachers and improves the ever ending the teacher and students relationships.

Teachers' Training

The teachers' training is again one of the essential components of the effective teaching because only well qualified with full subject knowledge may transfer good and recent knowledge of the subject to the students. The science is dynamic and its principles and concepts are changing very fast, so its teacher's duty to learn the recent development in science and technology for being good teacher, otherwise they may be laughing stock in the class rooms. Our new generation is very fast in learning, therefore, all teachers are required to be up to dated with subject knowledge so that they may satisfy the queries of the students. The NAARM. ICAR and some SAU's have developed teaching methodology and also providing the training to the teachers but these training have been linked with career promotion activities. Our senior teachers are hardly willing to take the training. They have their old subject notes and their teaching is also bases on these old notes. Certainly it is not very sound practice. Te teachers' training should be mandatory to all teachers for quality teaching. Our teachers should also be encouraged to visit abroad so that they may also learn the new practices of teaching, research and extension activities for quality improvement in our present system.

Finally, the 'teacher-students-teaching infrastructure-environment continuum' is to be maintained and improved for effective and quality teaching. The quality teaching of subject Agronomy may be stressed upon the highly practical oriented, interactive and experiential learning, need based and site specific research on project mode, inclusion of employment generating courses and building of satisfaction and confidence among students in the subject. Besides, extra curricular activities and teachers' training may add tonic in the teaching and learning process.

Chapter 24

Successful Teaching of Plant Pathology

Teaching is a process of guided interaction between the instructor, the student, and the subject matter. In Agricultural Universities, teaching is the single most important activity of the three functions-teaching, research, and extension. The last two are also undertaken by other agencies like research laboratories, Industries, state Department of Agriculture etc. Plant Pathology is an important subject as it has its bearing upon increase in quantity and improvement of quality of food. The quality of education in a department of university or college is judged through the performance of outgoing students. In order to produce well equipped technologically skillful, devoted and sincere human resources, it is essential to have quality education and to achieve all these good teachers are urgently needed to deliver the good.

Source of Plant Pathology Teachers

In Plant Pathology, the persons who are assigned the job of teaching the subject come from Agricultural Universities, Agriculture Colleges, Agriculture Institutes, and departments of Botany of non-agricultural universities and colleges. The extents to which these persons are exposed to the subject differ as the level and system of education is different in these institutions and the facilities to which have been exposed.

Areas of Training

It appears essential to train the persons, so called 'would be instructors' to acquaint them with the principles and philosophy of education. They need to be trained and exposed to teaching methods, lecture preparation and presentation, student attitudes-and psychology, adaptation to different system of imparting education in different institutions, use of teaching aids, arranging in conduct of practical classes and field trips, assignment preparation and presentation, involving students in discussion, collection of samples, herbarium preparation on diagnosis aspects setting questions, evaluation and grading,

Selection of Teaching Methodology

Bachelor's degree students generally come from higher secondary School or Intermediate Colleges after taking a basic course in Botany. The post-graduate students come from agricultural colleges, agricultural institutes, degree colleges and universities with a B.Sc. or B.Sc. Ag. degree for getting Master's in Plant Pathology. An instructor has to know each student closely about level of their knowledge and capacity to understand further. Teaching methodology is generally based on the number of students in a class and how much of what has to be taught. The instructor has increased responsibilities in the internal assessment system An instructor has to choose of lecture, question-answer, group discussion etc. based on need. The course content and the depth of content of each course need careful consideration. An instructor has to draw clear objectives for a course as to what the students are expected to do after completion of a particular course.

Students differ in their study habits. Recently, outgoing students were asked to classify the way they utilized their time at this university. Their responses are summarized below in hours per day.

A teacher has to arrange his lecture in a way that the ratio of class time to study time may be in the order of 1 to 2 for undergraduates and 1 to 3 for postgraduates. It is further evident from the above data that the students devote more of their time in the study of lecture notes and very little time in reading library books and other reference material. It becomes the responsibility of the instructors to plan the of his course in a way that the students depend more on guided library source. As the science of plant Pathology is developing rapidly, it is essential that the students are exposed to the recent literature by having them to write term papers even at the undergraduate level.

Hours per day

Study Habits	*Undergraduate*				
	Year				
	I	*II*	*III*	*IV*	*Ave.*
Class	4	4	6	8	5.5
Practical training	5	4	2	2	3.0
Study in hostel	3	2	1	2	2.0
Study in library	1	1	1	1	1.0
Other study	2	2	2	2	2.0
Sports, Co-curricular	1	3	2	1	2.0
Activities and recreation	1	1	2	1	1.5
Total	**17**	**17**	**17**	**17**	**17**

Further break up of the utilization of the study time is given below in percentage:

(%)

Purpose	*Undergraduate*				
	Year				
	I	*II*	*III*	*IV*	*Ave.*
Lecture notes	45	52	60	35	48
Text books	38	18	10	10	19
Library books	8	16	5	8	9
Cyclostyled material	0	3	10	20	8
Assignment Preparation	7	9	10	12	9
Other study	5	2	5	15	7

Teaching Aids

Teaching aids are of great help in making students understand a topic but an instructor needs proper knowledge and experience of their use in a most effective way. Examples of various kinds of teaching aids that may be used in plant Pathology course include:

- ☆ *Motion Pictures*: Life cycle of pathogens; working of plant protection equipment
- ☆ *Film strips*: Developmental stages of a pathogen or symptoms of plant diseases etc.
- ☆ *Slides*: Presenting data and also symptoms diagrammes, disease or life cycles etc.
- ☆ *Microphotographs*: Structures of pathogens
- ☆ *Over-head projector*: Presenting data or lecture outlines, diagrammes disease or life cycles.
- ☆ *Epidiascope*: Projecting printed material from books and bulletins
- ☆ *Model exhibits*: Charts are of great help in teaching on various pathological aspects.
- ☆ *Flannel graphs*: can be very effectively used to explain the life cycles of pathogens to undergraduates especially. In this method, the students remain curious to know what is next.
- ☆ *Auto-tutorial*: Instruction sets are good for use by students after the regular classes.
- ☆ *Tape-records*: as a means of improving their delivering of lectures some instructors have effectively used to record their lectures for voice quality, pronunciation, speed, clearity for subsequent replay and self evaluation.
- ☆ *Projectors/Electronic Media*: Such means have also been used for giving training to the students. Presentation and seminars through slides, transparency, power points and electronic media may also be used frequently. Students may also be advised to use CD-ROM/Internet to get more information on the subject. Video conferencing may also come up in near future on some special topics delivered by concerned experts.

However, the distribution of one page synopsis sheets for each lecture containing an introduction, breakup and outlines of the matter, reading assignment and model questions for discussion has proved to be of much help in motivation of the students to read books and reference material.

Practicals

Instructors have great opportunity for giving practical training in the laboratories and field-classes, they have to devote considerable time and efforts in planning, arranging and conduct of this part of their course training. Laboratory manuals containing a clear explanation for the purpose of each exercise, the equipments required, the procedures to be followed, the observations, and calculations to be made should be prepared. Instructors should know the best techniques to work with different organisms, time taken in their growth for display and stage to be used by the students during the practical class. They need to have proper knowledge of preparing specimen for herbarium and museum and their maintenance to transmit the same to the students.

Student's Evaluation

Evaluation of students depends entirely on the instructor. Preparing a good question paper is itself is both art and science. How the question need to be set to assess the grasping and retention power of the student. He should have good knowledge and experience with different types of examinations as well as questions to be set *i.e.* quizzes, hourly, mid-term, assignments and final practical and theory examinations and the type of questions to be included like brief-subjective, objective type, true-false gap-filling and multiple choice type. After evaluation the answer books are to be returned promptly to the students. The grading method must be just and uniform. Many instructors find a discussion period helpful at after shows valuated answer scripts.

Qualities of a Good Teacher

Lastly, a person who wishes to become a successful teacher may be given an opportunity, to know the qualities, which a good teacher may possess, so that by acquiring most of them he may do his job successfully:

- ☆ Loves for and profession
- ☆ Punctuality and sincerity not only in class-room but in general.

- ☆ Thorough knowledge and understanding of the subject.
- ☆ Enthusiasm about his subject and interest in subject matter delivery.
- ☆ Clarity of expression with appropriate voice.
- ☆ Questioning shall so that the students may put forth their best effort in discussion and learning.
- ☆ Encouragement and motivation of student.
- ☆ Use of need based visual aids
- ☆ Alert and perceptiveness.
- ☆ Knows budgeting of time skillfully.
- ☆ Fairness to students in examination and transparency in evaluation.

Course and Teachers' Evaluation

Each teacher would like to know his effort in covering a course/thesis, how the students felt about the course, its contents, the delivery of subject matter to the satisfaction of the student. A simple questionnaire may be distributed to all students comprising of 4-5 short questions to get feed-back from students where following few questions may be included on the last day of lecture/Lab.

- — Your teacher would like to know if there is something which he has done well, especially for your knowledge.
- — If some thing can be added/given more emphasis in the organization of course content.
- — That might be done to improve the teaching of this course.

Suggestions for Implementation of Teaching

The prime responsibility for improving the quality of teaching rests with the individual teachers and with their parent departments and institution. Much can be done at the department level through meetings, conference among the teaching staff, seminars and workshops on aspects of teaching and student evaluation. To supplement and extend such efforts, special institutes or workshops at a university/national/national level on teaching Plant Pathology should be organized. The services of some excellent teachers in the subject and education may be utilized for imparting training. A committee may be formed to finalize the whole programme including, course programme, course content, syllabus, period of training etc. The exchange of ideas, teaching materials, laboratory manuals, etc. among plant pathology teachers should be encouraged at a national and international level.

Chapter 25

Experiences in Teaching Plant Pathology

At the outset I would like to state a principle of Genetics in Biology that "as like begets like"; "good teacher begets good students" who in turn become good teachers and researchers. I had been fortunate enough for having been taught and guided mostly by good to excellent teachers in primary school, middle school, high school and in the university system of education to earn a final degree of Doctor of Philosophy in Plant Pathology from GBPUA&T, Pantnagar. On all the occasions at various levels of degree programmers I was made confident by my teachers (Prof. V.P.Gokhale and Prof. P.A. Shinde as teachers in B.Sc. (Ag) and M.Sc.(Ag) studies at Agriculture College Nagpur and Prof. Y.L.Nene as major Professor for Ph.D. studies as advisor at GBPUAT Pantnagar) that I have had made a right choice of studying agriculture as science in general and Plant Pathology in particular. I had lasting impression in my mind that society needed agriculture and agriculture needed Plant pathology.

It is worthwhile to mention herewith that in Europe in Anton de Bary's (a medical doctor who was later acclaimed to be father of Plant pathology) days that physicians (medical doctors) were primarily botanists because they depended during those days heavily on herbs to treat human diseases. Anton de Bary, medical doctor turned to become plant pathologist, himself was a great teacher and credited to have trained well known persons as world class Plant pathologists and teachers like Marshal Ward (discoverer of host-parasite interaction; coffee rust) and Millardet (discoverer of Bordeaux mixture as fungicide), the same is true with E.C.Stakman at the University of Minnesota who is credited to have trained Norman E. Borlaug (Father of Green Revolution and Nobel Peace Prize winner) as Plant pathologist; J.C.Walker Prof Plant pathology at the Univ. of Wisconsin who is credited to have trained plant pathologist like D.N.Srivastava; and K.C.Mehta, Professor of Botany, Agra College Agra, who is credited to have trained noted plant pathologists like L.M.Joshi and M.V.Rao in India. Most of the above were great teachers with strong sense of imparting teaching as profession with research experience in their respective field of specialization.

Experience as Teacher-cum-Researcher

Turning to my experiences as a teacher, I definitely may not come to the level of high degree of excellence at par with the above mentioned honourable personalities as teachers in Plant Pathology. But having given this opportunity on this occasion I would like to share my experiences as teacher as below.

Institution of an Examination on the First Day of the Class

My job as Assistant Prof/Associate Prof/Professor between 1972-2004 at GBPUAT Pantnagar entailed me 1/4th of full time to do research and the 3/4th of time to do teaching during the above period; *i.e.*, having a teaching load initially of five subjects (Diagnosis of Plant Diseases, Diseases of Field crops, Principles of plant pathology, Phytovirology, Introductory Plant pathology) and Seminar Class-Incharge spread over three trimesters initially and later over two semesters in an academic session. Later I was encouraged by the Department and Agriculture College faculty to develop a new course "Principles of Plant Disease Control in 1973-74. This great course was designed as Plant Protection/Plant pathology elective course for undergraduate students which was soon adopted by other Agricultural Universities as a model course and was required to be credited as core course by M.Sc.(Ag) students in Plant pathology and Agricultural chemicals besides being optional course for PG students in Entomology, Agronomy, Genetics and Plant Breeding and Horticulture disciplines. As a way of evaluating variable background of students, I usually instituted an examination (short quiz) on the first day of the class. This, of course, caused great consternation among students, even though I made it clear that examination was just for my information only as a teacher to know students background coming from different disciplines as they are being admitted from intermediate and B.Sc.(Biol) streams. Students seemed to be indoctrinated with the importance of examination that they were unbelieving when told otherwise. This practice had helped me to pay better attention towards students those who had come from math/physics and B.Sc. Biology background. Also some undergraduates had come from Hindi medium system of high school and Intermediate courses. Consequently special assignments were given to needy students and they were monitored by me very seriously to improve their performance in studying Plant Pathology courses in particular. Students were advised that they need to come to the expectations of fulfilling some social obligations to serve the society in general and the farmers of India in particular. *This was a rewarding experience as I had known from the feedback received from a quite a few number of pass out students.*

Term Paper Assignment

Another feature of the instruction was to assign a term paper (5-10 marks) on the latest developments in science of Plant pathology. It is noteworthy to mention herewith that quite a few students cleverly would like to get their term paper completed through their senior or junior best friends who could become evident to me by making out a difference in the handwriting of the bona fide registered students and otherwise. Thus idea of term paper writing was defeated and hence the same thing was then transformed into actual presentation besides asking students for submitting only hand-written (not allowing even typing of the matter) text for the purpose of evaluation. This practice was not only useful to an individual student but also to the whole class particularly during available time, if any, after completion of the scheduled practical exercise in the laboratory.

Class Administration

The act of conducting a class and imparting instructions with illustrations made on the black board or chart etc is a sort of administration confined to the classroom. Maintenance of punctuality of coming to the class in time, observing discipline, attendance and attention in the class were told to be prerequisites for students who registered the courses for which I used to be an instructor. This was quite useful to complete the particular class with great satisfaction allowing students to spare total five minutes time for asking questions both at the start and at the end of the class. Asking pertinent questions to students in between also had been my habit while dealing with a particular topic in the class. This had been greatly useful to detect hidden act of indiscipline particularly among undergraduate students. Once 6 U.G. students, after giving attendance to the roll call in Introductory Plant Pathology course, silently ran away (for preparing hourly for some other course) from the class when I turned away towards the blackboard for writing some points. I continued imparting instructions and as usual asked certain questions to a few students by name. But to my utter surprise a chain of 6 students whom I referred to by name were absent from the class. I then again took the attendance as per the roll call (keeping a watch on each student) at that particular moment and found that a few students had run away. I then took the whole class to the task. Each of the six students was termed by me as "notorious terrorists" and determined to report their cases not only to the University authorities but also to their parents informing them as "you are wasting money on your son/daughter and it is better that he/she is called back from Pantnagar as he/she is no more interested in studies". This matter was informed to these faulty students (one of them was a girl student) during lunch hour by their fellow colleagues on the same day and immediately in the afternoon (3.00PM) all the six came to my room in the Department and apologised and took the pledge of not repeating the same in future. I then insisted that this pledge should be given by them in writing.

The habit of asking questions to students by name has also been useful in keeping the students attention as I had otherwise experienced that some students do come to the class but remain busy at back benches (seats) reading or looking into their notes preparing for "hourly examination" of some other courses ignoring what is being taught or discussed in the scheduled class for which students make their presence for the sake of attendance.

On Technical Terms and Understanding Plant Pathology as Science

Teaching is an act of faith. One must believe in what one is doing, and that one can produce an appropriate answer to the question that has been asked. Students should be allowed to share their ideas with the teacher. For example, I was willing to speak frankly and critically when an occasion demanded. Preparing lectures was a demanding but excellent way to learn a subject in depth.

I was often distressed by what I regarded as improper or imprecise usage of technical terms in publication or by students in writing or verbally answering questions. The terms that distressed me the most were "pathogen", "virus" and "parasite" etc. *i.e.* the wrong usages of these terms by students. For example, students used to define pathogen very casually as "any agent that causes suffering" and virus as "an obligate parasite causing infection in plants". By cross questioning the answers, the students were carefully asked to answer whether "zinc" as deficient element in the plant is a pathogen as any agent that causes suffering is to be defined by them as pathogen. The answer then was clear "NO". Then what is pathogen? Thus any infectious agent is termed to be pathogen is said to be appropriately defined. Thus the moment some one says virus is a parasite; one is definitely take into account virus is an organism which is not true beyond any doubt. Hence appropriate definition of virus is to be understood as "an infectious nucleo-protein is a virus". Similarly whether disease is seed transmitted or pathogen is seed transmitted; the answer is pathogen is seed transmitted because development of the disease is a process. Understanding of vertical and horizontal resistance in plants against pathogen could be explained by vertical and horizontal mobility of a particular person in doing the assigned job as vertical mobility entails differential interaction and horizontal mobility refers to uniform interaction with people around the person elevated or transferred to similar job positions, respectively. Development of oilseed crop management strategies including developing a "Divya" mustard plant ideotype as a result of research at Pantnagar have been quite rewarding experiences in practically demonstrating the plant disease control strategies in courses like principles of plant pathology and Principles of plant disease control. These are only a few of many such examples which developed confidence in students in understanding the subject of Plant Pathology most appropriately.

Another aspect of teaching of any course by me was giving research paper reading assignment and its presentation by each one of the registered undergraduate students in "Introductory Plant Pathology" course during spare time, if any, of the practical class of the course. This is because I impressed upon the students that one should refuse to admit or advance any truth or fact until proved by strict technical experimental procedure. This had added confidence among students besides their regular practical experimental exercises particularly on seed treatment, spray of fungicides, host resistance and bio-agents concerning control of plant diseases. For post-graduate students the research paper reading assignment was given at least for 15 marks weightage exam (as III pre-final exam) which could be included in their marks for final evaluation. This has helped in a long way to students in understanding the subject or particular topic in more meaningful and rewarding manner.

I along with other faculty members [Dr.Y.L.Nene (as Head), Dr.S.P.S.Beniwal and late Dr. R.K.Tripathi] contributed notably to the stimulating environment for students in the Department of Plant Pathology. Students (Dr.U.S.Singh, Dr.V.S.Pundhir, Dr.J.Kumar, Dr.S.D.Sawant, Dr.K.D.Sharma, Dr.R.K.Jain, Dr. R.K.Khetrapal- to mention a few) were expected to work in the evening and we often came by to check on the students and lead discussion on various topics of Plant Pathology and related subjects like Biochemistry and Genetics and Plant Breeding including preparing students for professional maturity and academic preparations for oral examinations. We often then practiced in the Department rehearsal of preparation to be made at the meeting of Indian Phytopathological Society or Indian society of Mycology and Plant pathology. This has then consequently resulted in bagging several awards by students and staff of plant pathology in those days particularly in 1970s and early 1980s. In fact quite a few of us including myself did not recognize Sundays and Saturdays but spent hours arranging photographs and plates for publication. The University Library was of great help in those days, which was in fact a pride of Pantnagar; as the University Library was kept open for use of students/staff even during festival days upto midnight hours.

As Seminar-in-Charge Teacher

On several occasions in trimester as well as semester system of education I was assigned the job of being departmental Seminar Incharge for PG students. The seminar topics were to be assigned by the Seminar Incharge or by students themselves keeping in view the latest developments in Plant Pathology. I finally used to design the seminar class and served as a sort of advisor to students for preparation of their presentation and abstract on the seminar topic. My policy was to carefully edit each abstract or summary and when it was required I used to refer it back to the students major advisor for improvement or edit it myself. During this process students had a rewarding experience of understanding different terms related to seminar topic. As like other courses, I used to make attendance of registered students as compulsory and the same rules of attendance were made applicable as for any other course. Thus on some occasions because of shortage of attendance (beyond condonation) a few students were required to repeat the seminar in the subsequent semester. Some of the best students having the OGPA of 5 out of 5 could not deliver their seminar as per the expected norms (because of poor communication/presentation) and consequently such students could even earn only "C" grade despite being all "A"s in other courses. This had been shocking and 'news' like item not only to the students themselves but also to the faculty in the Department. But everything was taken in the right and positive spirit by the students for never ending scope for more and more improvement as a quality of students as products of the University education for fulfilling the social obligations.

Section 5
Communication Skills for Effective Teaching

Chapter 26

Creating Motivation in Class Room: Some Tips

Motivation is key to increasing work performance. Human behaviour is purposive and goal directed. There is motive behind all actions. Intense realization of needs arouses energized action. Class rooms should not be dull and drab, as they usually are learning should be naturally flowing and liberative rather than burdensome. Thus, there is need to work towards improving class room morale and motivate both teachers and students. Morale is high when both students and teachers approach the hour with interest and expectation. The morale of the class is low when little enthusiasm exists on both side of the desk. Above all, learning must be fun-entertaining and attractive. It is challenge that every teacher has to take to innovate approaches to create conducive climate for learning.

Motivation is critical learning. We need pay attention to motivational as pacts of learning, to tap the human potential. Learning must be fun a pleasant experience. Students should feel the urge to attend classes to gain the unique experience unfortunately classrooms are boring. Students leave classroom on smallest pretense. They want to escape from the dull and routine tranjectional. It is quite challenging for a teacher to remain in business without motivating the clients.

What is Motivation?

Motivation refers to the inner urge to take an initiative. All human actions are purposive in nature. No, rational person does anything without reason. Recall the times you feel attracted to a person; your attention reasons for motivation may be inside the person or solve a mathematical problem or draw picture or shoot Filenes just because he like it, feels satisfaction out of it else the rewards in physical terms money, position, power and facilities may be the external reason.

I recall how l used to feel motivated in the class of public speaking course because I had to opportunity to speak and express which I would not do otherwise. I could spend tire where as some of my classmates would find similar works boring and unnecessary. Thus, creating motivation has something to do with individual preferences skills and habits. However, there are some common and relationship which motivate most people.

Motivation of Students

Behaviour does not come out of vacuum. Motivation refers to arousal, sustenance and direction of behaviour. It explains what compels people to behave as they do. Human beings have different types of needs. Such needs initiate and sustain certain behaviour.

The students as human beings also have various reasons to behave the way they do in the class. They may get motivation to learn if they get opportunities for trying out new things, listening to a topic relevant to their everyday life, getting praise or recognition in the class or showing abilities in things they are good at. Thus, each student has varying degrees of need for affection, influence, recognition, new experience, etc. Besides, physical environment *i.e.* class room arrangements, furniture, light, teaching aids also stimulate interests of students learning process is hampered if the classes are dark, dirty, full of smoke with broken furniture or laboratory equipments. The place of learning must be attractive. Above all enthusiasm and manners of teachers also count a lot.Sometimes. students' own interest in the subject overcomes all the barriers and involves him/her in the class

Teacher's Attitudes

Teacher as facilitator, guide and motivator, is an important person in the class room. His personality and approach to content as well as clientele matter in evoking desirable response. However, teachers are known for their varying styles. It is a common knowledge that many teachers believe that is a better to keep students at a distance. Thus, they keep aloof and deal with students at impersonal level. They do not exchange introductions or encourage questions or interactions at personal level. They think they are right to behave in detached manner. There is little enthusiasm in such teachers. Their classes are lifeless and spiritless. The reward of good teaching, in fact, comes from the satisfaction of being able to exert influence on the lives and career of young men and women. Of course, there are instances that some teachers are extra strict. They always emphasize on rules and regulations. They consider students as inferior beings and enforce unusual standards of behaviour. Students should be taken as junior partner in common enterprise of teaching and learning. Students can also make powerful impact upon teachers through their questions, interests and novel approach to explore problems.

It is also common to see teachers who overemphasize on a particular text book or syllabus and have little attention on student's level, curiosity and experience. This created ennui. Unorganized, random teaching makes learning difficult. Teaching to be interesting requires enthusiasm on the part of teacher to treat his subject and clients well. A motivated and systematic teacher will, no doubt, create positive climate. Teacher's dress, manners and communication play important role in increasing class room morale.

Some Mantras for Motivation

Based on personal experiences and survey of literature, there are a number of ways to motivate students, as follows.

Relationship and Respect

Knowing students personally by name and their talents creates special bonds. As a teacher of communication, I have always found it helpful to know their names, backgrounds and even special interests. Students drop in my office and discuss the subject career. I encourage problem regarding communication and career. I encourage them to come. I have often found this motivates them to the class and course. Respect them as individuals by recognizing their positive attributes and contributions.

Introduce Some Victuals for Students' Involvement

Involvement motivates students. Students may be invited by turn, or at random to recapitulate the last class in the beginning of the class. They may be encourage individually or in group to summarize the topic covered today. Students may be assigned to bring certain observations/facts related with the topic and share in the class.

Use Group Discussion

Often lectures are heavy because of passive listening by the learners. There is no way to know if students understand or find it difficult. Thus, to break the ice, smaller classes (up to 35) can be quickly turned into groups of seven each without disturbing the physical setting. Give each group/row of seven an is we related with the topic.

Example

In a class on principles of extension education. After teaching the principles distribute each group a different problem of real extension work which needs application of one of the principles of extension. This generates discussion and helps them think. Various arguments for and against. Finally each group presents the answer which is explained by me.

Use Role Play

Using role play in class brings an element of drama and fun. I was teaching rural development problems and on reaching class I assigned roles of government officer, bank officer and farmer for disbursement of loan to fond the small farmer for new enterprise. They have to enact how farmers approach and how they are treated. This generates a lot of emotional involvement. At the end, it is easy to list problem rural poor face and why development programmes do not benefit

A teacher of genetic told me that he assigns roles of different geneticist to PG students, who enact the roles of farmers, scientists and defend their themes. The concerned details are distributed prefer hand.

Give Creative Assignments

Assignment can be given mechanically. But when you match

the assignments with interest and ability of students, it may stimulate them to work much move beyond your expectation. Meeting with students in small groups students normally do not open up in large classes. If you encourage them to sit in small groups of its by appointment to discuss.

☆ Putting hand-outs and course materials on line/in digital form

Create Relevance

Make students realize the importance. Convince them the need for change through new set of knowledge and skills. Discussion about career opportunities and interesting aspects may help them understand value provide opportunities for practice.

It takes time and lot of efforts to build new skills. Teaching communication skills is such a challenge. People always like to fall bock in troupes of old habits. Continuous practice and sincere feed back for a long time gulps people make the transition. More over, people with less prior exposure need more time. Build practice session beyond class hours. I recall how classes on video production would be conducted beyond class hours. After teaching the student's theory and come practical tips, they were divided into groups and allotted time according to their convenience to learn in off hours with help and support from instructor.

Bringing a Guest/Resource Person in Classroom

A change of face or person with particular experience can lift the spirits of students. I often bring my ex-students (who are in job or specialized studies) to my under-graduate class in extension education to share their experiences in the field and classroom. This helps the students to understand and appreciate the course. Sometimes another teacher with particular field experience may make the subject lively. Most often distinguished teachers from outside visit for examination and other purposes. They may be specially invited to share some useful thoughts to boost the morale of the class.

Well organized field visit to sites related with course if field visits are planned ahead of time and on the me directly related with the course, it can motivate students. Prior orientation to students about observations to be taken and discussion on the experience may be helpful. Keep team assignments, games and elements of cooperates in the An occasional but interesting quiz.

- ☆ Taken reward or appreciation
- ☆ Assigned teaching by students
- ☆ Mee

Some More Techniques to Motivate Students

Teachers must put efforts to create a desire among students to learn. Students should feel welcome and discover new facts with pleasant experiences. Here are some suggestions for teachers to create positive expectations.

1. *Enhance curiosity*: It is very interesting to create a sense of discovery or exploration by use of appropriate problems, case or pointed questions. Use of audio-visual aids can also create interest.
2. *Encourage creativity*: Every student has certain degree of originality in solving problems or perceiving concepts. Teacher can provide opportunities to students to use their creative abilities. It depends on teachers to devise assignments.
3. *Use group work*: Students prefer to be in group. Thus, tasks that use group work can stimulate cooperation among members. This also provides simulation of work environment in which they have to work I group.
4. *Give praise*: All students have desire for recognition and support. Teacher's attention and words of praise stimulate their interest.
5. *Provide positive feed back*: Students have desire to progress from one step to another. It is teacher's duty to provide his positive feed back and encourage to try for higher goals. You do not need to lie. You must tell what good you have observed and how improvement can be made. It is the job of a good Counsellor to help in improving performance through, feedback, motivation and helpful advice.
6. *Encourage competition*: Youths have a spirit of adventure. They want challenges. A healthy sense of competition to excel in a project or activity may inspire interest.

7. *Provide lot of active participation*: There is no substitute to practical experience. Where ever possible learners should get opportunities to do things on their own to experience the problems and gain confidence.
8. *Provide lot of variety*: Sitting in a class room for hours is boring. Variety in instructional activity lecture, team work, individual assignments reading or debate may infuse new vigour.
9. *Provide time to understand and practice*: Students take time to absorb new materials. Thus, enough time should be kept so that they can absorb and use at their own pace.
10. *Let learners make choices*: Teachers should create situation in which students can opt for things of interests for project, assignment or group work. Their involvement in decision making will enhance participation. Let the students chose topics for presentation or assignment according to their interest and put creativity. In a class of extension education students were asked to collect data through different techniques of data collection by choice. One students chose case study and made her study so lively,creative and meaningful that she felt she has been totally involved in the task like an artist.
11. Hold orientation session at the start of the course to know students personally and clarify expectations and significance of the course.
12. Introduce rituals to involve students in the beginning, middle and end of the class.
13. Organize group discussions on topics that need to be expanded in light of experiences/reading
14. Provide hand-outs or gist of every class in some form
15. Provide plenty of examples analogies and explanations to make the text understandable.
16. Organize your lectures and laboratory sessions in a manner that theory is followed by practical.
17. Make sure that students understand relevance of course early in the course
18. Plan assignments to stimulate interests of students.

19. Recognize and appreciate active learners.
20. Bring in interesting resource persons in classroom to share experiences and interest.
21. Encourage students to meet after the learn and clarify doubts.
22. Let learners choose activities and assignments out of their interest.

Motivation is critical to learning. It is also critical to teachers attitude towards students. We can not ignore it. Efforts must be made to catalyse interest to enhance learning.

Chapter 27

Personality Development of Teachers

Teachers are like leaders. They influence a generation or two. It has been a common knowledge that students have recalled the importance of teachers' personality on their achievements. Even otherwise, a teacher is a leader in the classroom. He teaches human behaviour by actually managing the situation. He motivates, guides, coaches and impacts the lives. A teacher is a nurture like a gardener. Thus, he must continuously learn to upgrade himself personally and professionally. There has been little work on this in past especially in agricultural higher education. Like management development programme of manager, regular teacher development programmes should be started for teachers to enhance professionalism. Personality should form an important part of the curriculum. Those who influence personality of students must themselves understand about their own. Self-concept, self-esteem and self disclosure are important components in this besides, communication skills, decision making, leadership, negotiation skills, conflict resolution, etc. The concept of personality of person embodies the "whole" of the being and reflects his/her actions, reactions and interactions with others. Usually it is best described by the traits which the individual exhibits. Personality has been defined as "the dynamic organization within the individual of those psychological systems that determines his unique adjustment to his environment."

Determinants of Personality

Hereditary

Studies and evidences suggest that genes affect both physical form and psychological traits. Studies of identical twins separated at birth and raised separately show startling similarities in tastes and interests.

Environment

An important determinant for personality development which includes culture, conditioning,norms, attitude and values

Situation

This is the third factor which also influences the effects of the other two factors on personality.

- ☆ Different situations differently affect behaviour.
- ☆ Situations impose varying constraints on behaviour.

Personality Traits

Traits are relatively permanent characteristics of an individual that describe his/her behaviour. *e.g.*, aggressive, timid, shy, lazy etc. Traits are many and the ones that are core or primary traits are not easy to identify and agree upon. Studies have identified hundreds of traits, but now a generally accepted 16 primary traits are in prevalence.

Sixteen Primary Traits

1. Reserved	vs	Outgoing
2. Less intelligent	vs	More intelligent
3. Affected by feelings	vs	Emotionally stable
4. Submissive	vs	Dominant
5. Serious	vs	Happy-go-lucky
6. Expedient	vs	Conscientious
7. Timid	vs	Venturesome
8. Tough-minded	vs	Sensitive

9.	Trusting	vs	Suspicious
10.	Practical	vs	Imaginative
11.	Forthright	vs	Shrewd
12.	Self-assessed	vs	Apprehensive
13.	Conservative	vs	Experimenting
14.	Group dependent	vs	Self-sufficient
15.	Uncontrolled	vs	Controlled
16.	Relaxed	vs	Tense

Self Awareness

Journey of a thousand miles begins with a single step-In the journey of life you must begin fro self. Aristotle rightly said, "know thyself". Every person has an image about himself. It may be a general foggy image of self as active, smart, dull, affects our day-to-day living. It has constant influence on our perceptions of situations, people and self. It is easy to tell about others and very difficult to point out critically our own strengths and weaknesses. How can we discern true self? We analyze behaviour through observation. Visible signs of happiness, anger, involvement distractions etc. indicate reality. Thoughts and feelings are the expressions of inner self. Deeper investigation of one's own feelings, beliefs, behaviour does not guarantee change by knowing only. Real change comes only when we make efforts to do something about the image. However sensitivity to the images and beliefs deeply entrenched in our minds reflects our gaps and stimulates action in the right direction. It is a mental picture of we think we are. It forms the core of our being and guides all our behaviour. There is magnetic power in what we hold as true.

Sources of Self-Concept

Early Childhood

As children we are not able to critically examine the experience around us. We go through the experience and have absorbed the feelings, thoughts and habits. Early childhood experiences and nurturing have deep grooves in our mind. We associate our self concept with early past experiences and think and decode accordingly. It is very difficult to guess the reasons behind wealthy person with humble roots.

Significant Others

Those who have controlled our lives during early days do influence with their attitudes and decisions. Out of our close proximity and respect as well as emotional attachments, we perceive self according to their reactions towards us.Friends, spouse and peers close to us have power to influence our self perception.

Major Life Experiences

Events or occurrences in our life that force us to reexamine our perceptions of self do bring radical changes in our self perception. We are aware of near total changes in lives of people after chance accident, brush with death or encounter with certain persons.

Self-Beliefs

Finally our strongly held beliefs are the filters through which we see ourselves and they influence our behaviour considerably.

The proverb that says 'you are what you think you are' emphasizes the value of self- concept. Positive self confidence provides strength to put efforts and succeed. Positively primed people are enthusiastic about what they choose to do and expect suckers. The power of thinking affects others/external situations to attract positive results. Life is like a mirror, you get what you give. What you give is decided by what you think. Some people always think in extremes. I am introvert, I am shy, I can make it in public. I can never succeed kin public jobs; such extreme thinking leads them to expect little of self.

A Word About Ideal Self

People have illusions about self and wrong expectations from life. Too much emphasis on perfection or idealism also inhibits people to asses self-rationally.

- ☆ The ideal self represents the aspirations and values of a person.
- ☆ One should differentiate between what one wants and what should or ideals have been imposed on self.
- ☆ Self-chosen goals and goals, considered on the basis of self-awareness are healthy.

Positive Self-Development

All of us remain engage in thinking. We talk to ourselves continuously. We engage in series of self- talk over matters that are important to us. The decision to do a thing is seldom taken instantly. If we fail in something significant, we look at it as a total failure and feel as if we can never make it happen. Such negative thinking affects our efficiency as well. Thus, it is essential that we keep our mind motivated through constructive thoughts. We should use positive thoughts to extricate negative thoughts. Take a break from such thinking and try to see success or failure in perspective.

Self-Esteem

Personal assessments of self and perception of worth or value give us high or low self-esteem. Self-esteem is a matter of standard set by ourself as ideal or worth achieving. Our acceptances in social groups also affect self-esteem. Experiences of success or failure also influence self-esteem. Much depends on our manner of comparing our worth and the standard we set. High self esteem gives us confidence to face competitions. Take risks, set high standards and be socially active.

Outcomes of High Self-Esteem (Khera,1998)

- Builds strong conviction.
- Creates too better relationship and fulfilling life.
- Makes a person more sensitive to others' needs and develop a caring attitude.
- Make a person self-motivated and ambitious.
- Makes a person open to new opportunities and challenges.
- Improves performance and increases risk-taking abilities.
- Helps a person give and receive both criticism and complements tactfully and easily.

Outcome of Low Self-Esteem

- Talking about people
- Critical attitude
- Arrogance
- Rebellious against authority
- Too much concerned about reputation

Change Your Self-Concept

Self-concept is the core of your being. You can choose to analysis your self-concept and change. Human being can achieve what they can believe and imagine. Continuous thinking and visualization help in actualizing the things of desire. You become what you except. Belief in self and acceptance for self are key to the positive results.

You can change your self-concept by changing your thought. Shakespeare said that nothing is good or bad but thinking makes it so. Thoughts which are continuously entertained in mind have power to turn into reality. Repeated thoughts with emotional feelings are more powerful. Watch your thoughts because they will become your words; watch your words for they will become your action. Watch your actions for they will become your habits. Watch your habit as they will become your character. Develop your character for it becomes your destiny. Thus, we should become careful about what we think.

Use Positive Self Talk

1. All of us engage in self-talk.
2. Often we think negatively about our performance.
3. We tend lower or exaggerate our weaknesses depending on our emotions.
4. Unrealistic expectations tend to influence our confidence.
5. Think of positive things about self.
6. Every day, choose to focus on your strengths and achievements.
7. Think and tell to yourself about the specific behaviour that leads you to distinctions.
8. Bring emotions into play. Think of love and appreciation of someone, you care most.
9. Concentrate on the thoughts which energises you. Repeat specific message in mind again and again which gives you peace and strength.

Use Visualization Techniques

Of the many gifts blessed to us by god, imagination is most powerful. If you can imagine something you can actualize it too.

Most of the inventions and achievements have been thought first, by the inventors. They had visualized them precisely in mind and seen them in dreams. They felt the power of the idea. Such passionate dreams kept them working toward the goals despite all odds. Edison, an ordinary book binder and peddler, dreamt of electric bulb and felt sure that it can be created. Same is true for those who thought of wonderful aeroplane, television, and computer and so on. Visualization has power. Visualizing things with belief produces results. sub-conscious mind obeys what conscious mind strongly believes.

Some Suggestions

1. Relax and perform breathing exercise.
2. Focus on your behaviour.
3. See specific change.
4. Mix special feelings with achievement.
5. Enjoy the success as if you've got it.
6. Concentrate for sure success.

Harmonious Blend of Conscious and Sub-Conscious Mind

Whatever your powerful mind believes very intensely would instantly come to pass. Thought has force as electricity or gravity...

Lahiri Mahasay
(Autobiography of a yogi)

Conscious mind is active usually during our waking hours. It thinks logically and is limited by the perception of five senses. It is always on alert to decide how to, in a given situation. Sub-conscious mind acts actively when mind is quiet and relaxed. Moments before sleeping or just waking up are ideal for receiving messages. It works wonders if mind is clear of any fear, doubt or thoughts and feelings and translates into reality. Whatever you can visualize during the relaxed phase without any doubt, you may hope to see them in reality. Deep conviction within, activates the process of actualizing desired results. It acts like negative image cleared be the positives. Sub-conscious mind is creative. It seeks no logic. If you visualize with conviction, you'd get it. Dominant thoughts with feeling produce wonderful results. You can fashion, mold or bring alive your dreams. You can be producer, director and actor of the film about you. All inventors, entrepreneurs, artists had magnetized their lives towards ideals they held. They beloved in the tremendous benefit of their great works for the humanity, persistent imagination with faith helped them overcome the barrier.

Communication Skills for Self Development

Meaning of Communication

The word communication originates from the word 'communis'; which means common. So communication is an act by which a person shares the knowledge, feelings, ideas, information, etc. in ways such that each gains a common understanding of the meaning, intent and use of the message. It can be generalized that communication is a process of social interaction, *i.e.*, in a communication situation two or more individuals interact. They try to tangibly influence the ideas; attitudes, knowledge, ideas or feelings taking place between two individuals. In a face-to-face situation communication is not a mere exchange of information but something more, because in such a situation, along with the information one passes, the gestures, expression, language, the manner of expression and tone–all these combined together, create a sort of impact on both. Some kind of change occurs as a result of interaction. This change may be visible in terms of knowledge and behaviour.

Communication can occur even without words. Our four senses–audio, visual, touch and smell communicate. The ringing of an alarm clock communicates to us that we are to get up, the peep outside the window *i.e.*, the visual sense, gives us the indication of weather–rain, snow, fog, storm, etc; the touch of a pot, whether hot or cold, communicates to us whether it can be handled or not; the sense of smell gives us knowledge whether meat or eggs are being cooked for breakfast. There are various ways in which people communicate; *e.g.*, the exchange of greetings differs from country to country. It is different even among different religions and groups of people in the same country. So a good communicator is able to find out feelings or reaction through varieties of ways.

Importance of Communication

Communication is an important part of human life. All living beings communicate. Animals and birds communicate too. Human communication is refined and lifelong process. The moment a child is born, it cries. If there is no communication at all, we suspect if the person is dead. Human communication is purposive. We do not even wink our eyes without a purpose. Understanding and using communication purposefully enhances our effectiveness in personal and professional lives. Urge to communicate is universal and natural. We want to talk, share and interact with others. Imagine how do we feel in a strange place with nobody to talk to. Professional success demands more skillful use of communication. Teachers need to motivate their students and instruct clearly. We have all suffered in the hand of teachers who could not communicate clearly. Politicians, salesmen, managers, actors, advocates, advertisers, public relation experts and others use communication for effective functioning. Salesmen try to attract attention, build interest, create desire and convince clients to arouse action. There are many industries thriving on communication for profits. Information is power and communication of information is big business as evident through astounding growth in printing, advertising, television channels, films, computers, information kiosks etc.

Communication is not merely speaking. We communicate much before we open our mouths. Studies indicate that people make impressions about us in first thirty seconds of meeting. Such impressions have indelible effect. Everything about us- our dress, gestures, postures, voice, mannerism, etc. convey something about us. Even if we are silent still we communicate. This is intrapersonal communication or self-talk. Much of what we do and how we feel is product of this self-talk. Self-talk shapes our decisions and actions. We talk with others on one to one basis. Conversation between two people is interpersonal communication. This helps us in building relationship, contacts, and network for personal and professional achievements. We also need to be in groups and interact. Communication in group is different than one to one communication. Some people feel comfortable in interpersonal situations but feel lost while in groups. Group communication is another interesting aspect of our social life. We have to work in groups and influence to rise higher ladders of success. We have to address relatively large student to inform, motivate or influence the opinion of the listeners. This is a leadership skill needed by people in positions of importance so that they can influence others. Public speaking is powerful medium to reach masses.

Public Speaking Tips for Effective Teaching

Teachers have the basic responsibility to deliver lectures and their effectiveness largely depends on their speaking skills. It has been observed that many very knowledgeable and learned teachers cut a poor image in classes due to lack of speaking skills. Few tips are shared to emerge as a good speaker though it depends on formal training which is available in ET Cell of the University. Feel free to share your problems with us regarding this.

Basic Guidelines

Be Totally Engaged in What You are Saying

Great speeches are not given by people who are thinking about themselves and how they look or sound. Pick the topic of the lecture and throw yourself into it with passion. Total engagement is directly proportional to less nervousness.

Use Cue Cards

Memorizing usually produces a stiff, wooden presentation and lacks human touch. Dictating doesn't work, because you lose eye contact with your student. Instead, put the key ideas, quotations, statistics, etc., on note cards. Then practice it mentally to have hold over the flow.

Don't deliver the lecture exactly the same way each time, but you can reinvent the style everytime.

Begin with an Opening that Catches the Student's Interest

Normally, we begin with-" I'm going to talk about...", or "My topic for today is...". These have proved to be the worst and redundant beginnings inhibits creative thinking and leads to monotony. Why not to change our beginning in each class which may vary from a quote to a puzzle to a case to an incident to some striking facts to a question and so on.

Pace

Speak as slowly as you can; then reduce your pace by half. Then add variety. Why? Two reasons. First, your nervous system will be running at high speed, and if you're not careful your ten-minute presentation will be over in three minutes—and no one will have understood a single word.

Second, you must remember that your student has to take in what you are saying through their ears alone. Go slowly, or else they will miss too much, and you will not be understood. Anything that doesn't change tends to become monotonous. Keep your student's interest and attention by varying your pace from time to time.

Volume

You must speak much louder than you normally do, without shouting. The sound should come from your diaphragm, not your throat. Vary your volume to retain your student's attention and add dramatic impact.

Eye Contact

If you don't make eye contact with your student, you will lose them. Don't stare at the ceiling, or out the window. Look at individuals, but don't look at just one or two people—move your eye contact around the room. If direct eye contact makes you lose focus, try looking at foreheads or hair instead.

Hands, etc.

Use them to emphasise your points, in a natural way, but don't move them around in a way that will distract attention from what you are saying, or put them in your pockets. Also: Stand up, without leaning on the wall, against a desk, etc.

Signal your Closing

A train slows down as it comes into the station. You should do the same. If you stop abruptly, you will confuse your student and lead them to wonder whether you have finished, or just lost your voice.

Return to the Beginning

If you begin with a quotation, a question, an anecdote, a startling statement, etc., then return to it at the end to complete it or comment on it. This ending provides a very nice "frame", a set of bookends to enclose your speech and give it structure. It also helps to signal your closing.

Maximizing Verbal/Nonverbal Delivery

Once you have your anxiety working for you (and not against you!), you'll want to consider the delivery style that will best complement your communication goal of engaging the student.

Verbal Delivery

Verbal delivery enhances understanding of your message. You'll want to work towards an elevated, conversational style. The students want you to talk to them not at them. Use a sincere tone that lets the student know you're invested in the material you're presenting.

- ☆ Incorporate pauses to highlight key ideas, transition between points, and/or grab the student's attention again.
- ☆ Think about the volume of your voice. Can the person in the back hear you clearly? Keep in mind, working with a microphone may make it necessary for you to modify your volume.
- ☆ Pay special attention to the rate of speech you're using; remember the goal isn't to finish as soon as possible. You want to make sure that the student is understanding the material and connecting with it. The rate you use should allow you to focus on explaining the material carefully.
- ☆ Having variety in your rate, volume, and inflection can help in giving emphasis on the material and sustain student interest. Be especially conscious of these vocal delivery features when giving examples or sharing an illustration or story.

Nonverbal Delivery

Nonverbal delivery enhances understanding of your message. Eye contact is crucial in keeping you and the student connected. The student forms understanding and makes judgments about your material based on their ability to see your eyes. You can see if you are making sense to the student by looking at them and gauging their nonverbal feedback.

- ☆ Anxiety can cause a speaker to see students as hostile menace rather than a gathering of interested individuals.
- ☆ To help you connect on a person-to-person level, locate the "friendly" faces in the student—the people who give

positive feedback in the form of smiles, nods and attentive listening positions. Speak directly to these people, varying your eye contact between them. You'll begin to notice more and more "friendly" faces to connect with as the student is drawn into your speech.

Gestures and Stance

Gestures and stances should appear natural. They should add purpose and emphasis to the message. It is best to use gestures unconsciously so that you avoid drawing attention to them, but if nervous energy tends to make you freeze physically, then you may need to plan a few gestures until you become more comfortable.

- ✰ It's better to have minimal gestures than distracting or overdone gestures.
- ✰ Minimize distracting habits like fidgeting or swaying.
- ✰ To help channel this nervous energy, think about incorporating purposeful movement. Not only will this use some of the energy, it will also help to emphasize main points and keep the student engaged.
- ✰ One idea is to picture a triangle on the floor and plot out the points where you may move during the speech (taking care to minimize distracting pacing). Wherever you place your points, it's important that you begin and end your speech in the same place to let the student know that the speech is concluding.

Assertiveness Development

Assertiveness is the ability to express your feelings, opinions, beliefs, and needs directly, openly and honestly, while not violating the personal rights of others. It is the ability to declare yourself, who you are, what you think and feel. It is an active rather than a passive approach to others, and to life. Assertiveness in communication and social relationships involves openness, honesty, and firmness, all with appropriateness and flexibility. The assertive person is confident in a relaxed way, as well as free and spontaneous in social situations.

Why to be Assertive?

One of the most significant implications of learning assertiveness is that it increases the individual's freedom. No person should be subject to the domination, whims, and aggression of others. There is only one way to eliminate such restrictions to one's freedom, and that is by being appropriately assertive with those who would smother or dominate. People must be free to choose how to act. If they are unable to be assertive in a situation, they are not free to choose what to do, or even whether to be assertive or not. If they are able to be assertive, then they can choose to be assertive or not, and they are free. This can lead to greater emotional freedom in general. People who are very non-assertive, passive and inhibited often cannot freely express other emotions like tenderness and real affection. Learning to express justified annoyance and anger, and to assert one's rights in a firm and straightforward manner, makes it easier to relate to people in a friendly and caring way. With reduced anxiety and uncertainty in interpersonal situations, a person can be less defensive and afford to risk greater openness and honesty in emotional expression.

Assertiveness Increases the Chances of Our Needs Being Met

If we are able to tell people what we want without becoming aggressive they will be more likely to help us. Also, if they can't help us and we are able to accept that without becoming aggressive they will remain friends.

Assertiveness Allows Us to Remain in Control

We can decide for ourselves what we want to do and then seek out opportunities to do it–or to do something similar. It puts us back in the driving seat.

Assertiveness Brings Greater Self-Confidence

As we learn to take control and see what we can achieve our confidence increases. This in turn increases our feelings of self worth and self esteem. We begin to feel better–more effective.

Assertiveness Lets Us have Greater Confidence in Others

This is because it also helps others to state their needs and wants. By dealing honestly and fairly with them we encourage them to do the same with us.

Assertive People have More Friends

As we begin to treat people more fairly they begin to trust us, to like us and want to spend more time with us. We make friends who truly respect us instead of walking all over us (passivity) or fearing us (aggression).

Assertiveness Reduces Stress

The more in control we feel, the less stressed we feel. We don't need to worry about doing things we'd rather not. We don't have to let other people control us. Nor do we have to worry about trying to control other people. We have the power to choose our own destiny.

Assertive vs Aggressive or Passive Style

Aggressive Style

Typical examples of aggressive behavior are fighting, accusing, threatening, and generally stepping on people without regard for their feelings. The advantage of this kind of behavior is that people do not push the aggressive person around. The disadvantage is that people do not want to be around him or her.

Passive Style

A person is behaving passively when he lets others push him around, when he does not stand up for himself, and when he does what he is told, regardless of how he feels about it. The advantage of being passive is that you rarely experience direct rejection. The disadvantage is that you are taken advantage of, and you store up a heavy burden or resentment and anger.

Assertive Style

A person is behaving assertively when he stands up for himself, expresses his true feelings, and does not let others take advantage of him. At the same time, he is considerate of others' feelings. The advantage of being assertive is that you get what you want, usually without making others mad. If you are assertive, you can act in your own best interest and not feel guilty or wrong about it. Meekness and withdrawal, attack and blame are no longer needed with the mastery of assertive behavior. They are seen for what they are–sadly inadequate strategies of escape that create more pain and stress than they prevent. Before you can achieve assertive behavior you must face the fact that the passive and aggressive styles have often failed to get you what you want.

Traits of Assertive People

1. Allow others to complete their thoughts before we speak.
2. Stand up for the position that matches our feelings or the evidence.
3. Make our own decisions based on what we think is right.
4. Look to friendships as opportunities to learn more about ourselves and others and to share ideas.
5. Spontaneously and naturally enter into conversations using a moderate tone and volume of voice.
6. Try to understand the feelings of other before describing our own.
7. Try to avoid harm and inconvenience by talking out our problems before they occur or finding rational means for coping with unavoidable harm or inconvenience.
8. Face problems and decisions squarely.
9. Consider ourselves strong and capable, but generally equal to most other people.
10. Face responsibility with respect to our situation, needs, and rights

Traits of Non-Assertive People

1. Stay in the back of groups.
2. Always stick to the middle-of-the-road position or refrain from taking a stand.
3. Allow others to make decisions for them.
4. Pass by potential friendships because they seem like too much effort.
5. Always keep our voice low or avoid eye contact to keep from calling attention to ourselves.
6. Verbally agree with others despite our real feelings.
7. Bring harm or inconvenience to ourselves to avoid harming or causing inconvenience to others.
8. Procrastinate to avoid problems and to keep from making decisions.
9. Always consider ourselves weaker and less capable than others.

10. Always escape responsibility with excuses and "good" reasons.

Basic Fallacies we owe that Checks Assertiveness

1. It is selfish to think of your needs first before others' needs.
2. It is shameful to make mistakes.
3. If you can't convince others that your feelings are reasonable, then they must be wrong, or maybe you are going crazy.
4. You should respect the views of others, especially if they are in a position of authority. Keep your differences of opinion to yourself.
5. You should always try to be logical and consistent.
6. You should be flexible and adjusting at all cost.
7. You should never interrupt people. Asking questions reveals your stupidity to others.
8. You shouldn't take up others' valuable time with your problems.
9. People don't want to hear that you feel bad, so keep it to yourself.
10. When someone takes the time to give you advice, you should take it very seriously. How can you deny that!
11. Knowing that you did something well is its own reward. Don't tell it to others as it shows your show-off nature.
12. You should always try to accommodate others.
13. You should always have a good reason for what you feel and do.
14. It's always a good policy to stay on people's good side.

Your Legitimate Rights

1. You have the right to put yourself first sometimes.
2. You have the right to make mistakes.
3. You have the right to be the final judge of your feelings and accept them as legitimate.
4. You have the right to have your own opinions and convictions.

5. You have the right to change your mind or decide on a different course of action.
6. You have a right to protest unfair treatment or criticism.
7. You have a right to interrupt in order to ask for clarification.
8. You have a right to negotiate for change.
9. You have a right to ask for help or emotional support.
10. You have a right to feel and express pain.
11. You have a right to ignore the advice of others.
12. You have a right to receive formal recognition for your work and achievements.
13. You have a right to say "no."
14. You have a right to be alone, even if others would prefer your company.
15. You have a right not to have to justify yourself to others.
16. You have a right not to take responsibility for someone else's problem.
17. You have a right not to have to anticipate others' needs and wishes.
18. You have a right not to always worry about the goodwill of others.
19. You have a right to choose not to respond to a situation.

Leadership

The subject of leadership is ever green. Everyone wants to become a leader. The lust for power, position and prosperity attracts most human beings. That is why studies on leadership have wide public appeal. The call for leadership is eternal to all societies. There is always a need for people who can mobilize others. Where are such people with vision, vigor, enthusiasm and abilities to hold people together for a cause in every society ? There are the people who count, whose presence motivates, guides and controls activities of common interest. People want to listen and carry out their commands. Such individuals are called by many names, such as, elites, power centers, power holders, leaders. Leaders are essential part of any group.

Most of us associate leadership with domination, command or control. John Adair says that leadership is not a skill of dominating followers but of creating equals. It is not about how to create hierarchy but how to run them. The truth is that the qualities needed to lead are similar to the qualities needed to be a part of a team. In fact, in power structure of every society, there are vital individuals who promote, stimulate, guide or otherwise influence members for action. Leadership is a group phenomenon. In group, there are several members linked by need or purpose. Each member has influence over others. However, leaders have more influence over others than followers. In fact, both leaders and followers influence each other. The leaders affect by virtue of special characteristics or ability. The leader has influence. At the same time, followers' acceptance or rejection of leaders' ideas or action also affect leaders' behaviour.

Thus, it can be said that leadership can be understood in terms of its essential constituents, *viz.* leader, follower, goal and situation. Goal refers to the common purpose of the group. Situation is reflected through composition the group, nature of task facilities and work environment. Leader in one situation may not be leader in another situation because the challenges are different in different settings. It calls for different set of abilities in different settings. Leadership is essential part of group life. The essence of leadership concerns around the influence exerted by leader over the group. Leader initiates, motivates, guides and controls group activity. There is a mutually interdependent relationship between leader and followers. Leadership varies according to situation. Leadership ability exists in all the members to some extent. A teacher is a leader as he has to motivate the students, instruct them, manage resources and control.He guides a generation or two by his style of academic leadership.

Leadership Qualities for Teachers

Initiator

Being brave enough to reach beyond the boundaries created by our existing, often deeply held limitations, fears and beliefs. Initiating change in our lives-of any kind-happens only when we are courageous enough to take the necessary action.

Honest

Committing oneself to show up and be fully present in all aspects of life, removing the mask and becoming a real, vulnerable and intimate human being, a person without self-absorption who is genuine and emotionally and spiritually connected to others.

Dedicated

Focusing on the needs of others by listening to them, identifying their needs and meeting them, and thus inspiring them, as opposed to sticking with our usual self-focused, fear-based approach.

Truthful

Listening openly to the truth of others and refusing to compromise with integrity or to deny universal truths-even when avoiding the truth might, on the face of it, especially in testing times, seem easier.

Empathy

Embracing the underlying oneness with others and life. Relating to, and inspiring, others and touching their hearts in ways that add to who you both are as persons.

Effectiveness

Being capable of, or successful in, achieving the goals we desire in all aspects of life.

Section 6
Evaluation as a Tool of Teaching

Chapter 28

Evaluation in Teaching Learning Process

Education is a mean for empowerment and social development assumes much significance when agriculture is added to it. The purpose of agricultural education is to make a better professional in the field of agriculture and allied vocations and the whole of teaching-learning process revolves round the inter-relationship between the teacher and learner. It is known that an efficient teacher is one who aims at transferring knowledge and skills to the students to compete in the global job market and pursue a respectable living. A successful student is one who develops skills and an acceptable personality. But in last few years the scenario has changed. An efficient teacher is one who acquires a key position in the university administration and a successful student is one who scores good marks in the formal examination. Efforts to improve the quality of teaching depend for their effectiveness on the availability of accurate, detail and objective evaluation of teaching.

The success of the educational process depends so much on the character and ability of the teacher that in any plan of university reform the main concern is for securing an adequate staff with qualifications necessary for achieving the aims and objectives of the university teaching. Effective faculty assessment and evaluation systems are in the best interest of colleges and universities and their faculty. A strong assessment and evaluation system will lead to opportunities for faculty development that, in turn, will allow faculty to earn the greatest rewards for their contributions.

In the past, teacher evaluation has generally not been high-stakes activity, in part because improving the quality of teachers has not been seen as critical for improving the quality of education. Instead improvement efforts in the field of education were more focused on improving the curriculum, altering management methods and developing new programs. Now more attention is being devoted to teacher evaluation and its role in shaping teaching. Evaluation in teaching often referred as students evaluation by conducting various kinds of examinations. However the whole teaching-learning process involves four other components *viz.* teachers, subject matter, teaching material and the teaching environment which also needs to be evaluated to facilitate better learning among students. Keeping in view the importance of both students and teacher is essential to evaluate them time to time to make teaching-learning process more effective.

Teacher Evaluation

Teacher is a guide, friend, mentor and philosopher for students who play most important role in subject learning of the student as well as values of life. It is essential to evaluate teachers on certain aspects like command over subject matter, ways of presentation, language and use of modern instructional technology tools. The purpose of these evaluation are variously to provide, a source of diagnostic feedback to faculty about the effectiveness of their teaching, a measure of teaching effectiveness to be used in tenure/promotion decision and a source of information for students to use in the selection of course and instructors.

Teacher evaluation refers to periodic evaluation of teachers' performance. It involves a systematic gathering and analysis of information, on the basis of which decisions are taken regarding the effectiveness, efficiency and/or competence of the teachers in realizing set professional goals. Darling-Hammond and other (1983) define teacher evaluation as "collecting and using information to judge". It serves a variety of purposes which can be classified into two broad categories, namely

- *Formative purpose*: when results of such evaluation are used to improve classroom instruction, student learning and to foster professional growth of teacher.
- *Summative purpose*: when results of such evaluation are used for administrative/personnel decisions like promotion, salary increase, demotion, dismissal, awards and/or meeting public/government accountability demands.
- Efforts to improve the quality of instruction necessitate an accurate, detailed and objective evaluation of instruction so that teachers are provided suitable detailed feedback to improve in the desired direction. There are two main types of criteria on which evaluation of teaching can be based. These are evaluation of behavior of teacher while teaching and outcomes of teaching in terms of pupils' learning outcome.

According to Barrett, Joan teacher evaluation program consists of varying combination of following components.

Teacher Interview

This one-to-one conference is used to hire new teachers and communicate evaluation results to experienced teachers. An updated, formalized version, the Teacher Perceiver Interview, reduces possible interviewer bias.

Competency Testing

The National Eligibility Test (NET) is an example of competency testing.

Classroom Observation

This is the most popular evaluation method, usually performed annually by school administrators for experienced teachers and more frequently for beginning teachers. Observation reveals information about such things as teacher interaction and rapport with pupils that is unavailable from other sources.

Student Ratings

Using student ratings in teacher evaluation has been restricted to higher education, although student input has been collected informally in middle and secondary schools. This method is inexpensive, and has a high degree of reliability, but questions of validity and bias remain

Peer Review

Teaching colleagues observe each other's classroom and examine lesson plans, tests, and graded assignments. Peer review examines a wider scope of teaching activities than other methods. Disadvantages include time consumption and possible peer conflict.

Student Achievement

Nationally standardized student achievement examinations often are used to evaluate teachers and school systems by ranking the student, class, and school according to national norms.

Faculty Self-Evaluation:

This method usually supplements more formal evaluation methods and is used with other data to identify weak areas of instruction and classroom management skills. It serves as an important source of information for staff development, but is unsuitable for accountability decisions

Earlier it has been seen that either teachers are not evaluated or are evaluated by their heads in the form of Annual Assessment Reports. This may be biased or politicalised. As we know students are the best evaluators of teacher as they observe each and every activity of teacher in classroom. Since last 3-4 years an effort has been done by Pantnagar University. Teachers Evaluation Performa or rating scales has designed which are given to all students at the end of the semester. But these rating scales do not provide measures of the separate components of teaching effectiveness. It contains heterogeneous mixture of different items and students rating are summarized by an average of these items, then there is no basis for knowing what is being measured. Moreover students do not understand the importance of these rating scales and often fill it even without reading given statements. There is an urgent need to develop a whole system that will include instructors self evaluation, students rating scales, peers/colleague rating scales and evaluation by superiors/heads. In order to make evaluation transparent it should be displayed in notice boards or in university websites. This will also help administration to conduct teachers training at regular intervals.

Teachers' Evaluation by Students

It is one of the most controversial techniques used to identify teacher effectiveness. Students who have completed a course have observed teachers in classroom are in a position to provide potential useful information concerning the teacher's effectiveness. Several studies have been done to find out students' perception about effective teachers. **Miler Kahler and Rheault (1989)** identified five common performance areas for effective agricultural teachers: productive teaching behavior (which includes designing life like situations and activities), organized structured class management, positive interpersonal relationships, professional responsibilities (which includes completing duties in a timely manner), and personal personality traits such as humor and patience). According to **Luft and Thompson (1995)** students identified an effective agriculture teacher as having the following characteristics,; showing enthusiasm for teaching, serving as good role models for students, being committed to helping students learn, showing their commitment to teaching by belonging to professional teacher organizations, enjoying teaching, being self confident and poised, being prompt and on time and being neatly dressed and well groomed.

This technique help in continually sensitizing or reminding teachers that student are their customers. Students are not the evaluators; they simply provide data to the evaluators The key element is the good sense of the evaluators. Unfortunately, many evaluators have stereotype about what constitute good teaching, despite the fact that there are many ways to be effective. Thus they may undervalue a teacher because the students' judgment of their own learning may not fit with the pattern of rating on such characteristics as organization or enthusiasm, or other characteristics usually associated with effective teaching. Often the evaluators give less weight to the student ratings than to less dependable evidence, such as peer observation of teaching, testimonials, or general impressions of the teachers' personality. Student rating provide a means of communicating between students and instructor, which in large institutions, may not exist in other forms. Such communication may lead to the kind of involvement by student and instructor in the teaching-learning process that raises the whole level of instruction.

Few faculty members question the usefulness of rating in providing feedback about teaching that can result in improved instruction, but many continue to challenge student rating use in making personnel decisions **(Marsh and others, 1979).** A number of studies are there which attempted to find out the effect of gender, a variable presumably extraneous to evaluation of teacher performance, on student rating. **Bosow and Silberg in 1987** found in a study, a common although not universal pattern for female and male students to rate same sex instructors in some disciplines higher on some questions than other-sex instructors. Most reviews of students' evaluation research conclude that effects of gender variable are negligible.

The method of administering and gathering student response can determine the quality of the resulting data. If possible the responsibility should be given to instructional development or testing personnel. Regardless of who does the actual administration, a standard set of instruction and a designated time limit for filling out the questionnaire are essential. Informal administration can lead to bias in the ratings, low return of the forms, or less than candid response on the part of students.

Students Evaluation

The purpose of teaching is, of course, to produce pupil learning; it seems perfectly logical, therefore to evaluate teaching on the basis of outcomes. In most the agriculture colleges it is seen that examination system is not very strong as it tests only students power of retention of memory. Most of the students get good scores through one night fight before hourly or finals. Students are not guided by teachers to read textbooks or reference books because the teacher would like to suggest reading, in the present system, only those help books ad important questions which are only examination oriented. Even if students wants to learn the full course, some teachers are not prepared to cooperate because they want to teach selectively because of the large choice in the examination papers. On the whole the students are not allowed to see their checked answer books and evaluation is not done transparently, which sows incapability of teachers to evaluate students who don't trust. This lack of trust leads to loss of faith by students in themselves and it further leads to low performance by them. Students will not be alienated if they are fully satisfied by the transparent examination system. Following suggestion will make examination system more effective and trustworthy.

- ☆ More emphasis should be given to application based teaching learning and evaluation inside as well as outside of the four walls of the institution.
- ☆ Completely objective continuous internal transparent system should be introduce in which teacher examines the answer books of their own students.
- ☆ Some departments in Agriculture College have followed open book system. It should be followed in all agriculture colleges/institutions as it will discourage one night fight before examination.
- ☆ Term paper writing should be followed by actual presentation for the purpose of evaluation. This will also help students to improve their communication skill confidence.
- ☆ For post graduate students research paper reading assignments should be given which can be included in their hourly and final evaluation. This will certainly inculcate the habit of reading more and more research papers and articles on particular topics.
- ☆ Quick declaration of marks of examination with model answers will also help in learning.
- ☆ The idea should be to give marks not to cut the marks.
- ☆ There should be change in setting of question papers every years with respect to form and type of questions which will inculcate the habit of through study.

An effective students and teachers evaluation system will help in building good student-teacher relationship and they will understand their responsibility towards each other. No educational system can succeed if students and teachers are alienated from it.

Chapter 29

Designing Tests and Evaluation Approaches

Classroom tests play an important role in the evaluation of students' learning. They provide relevant measures of many important learning outcomes and direct evidence concerning others. The validity of the information they provide, however depends on the care that goes into planning and preparation of the test. The term test, measurement and evaluation are easily confused because all may be involved in a single process. We generally ask our students to answer a series of questions concerning our subject, obtain their scores by counting the number of correct answers and conclude that all the students are making good learning progress. Here, we are concerned with all the three concepts *i.e.* test, measurement and evaluation. A test is a set of questions, measurement is the assigning of numbers to the test results according to specific rule and the evaluation adds the value judgment.

The specific meaning of each term, as applied to classroom evaluation, is summarized as:

Test

An instrument or systematic procedure for measuring a sample of behaviour. (Answers the question, "how well" does the individual perform – either in comparison with others or in comparison with a domain of performance tasks)

Measurement

The process of obtaining a numerical description of the degree to which an individual possesses a particular characteristics (Answer the question "How much?")

Evaluation

The systematic process of collecting, analyzing and interpreting information to determine the extent to which students are achieving the instructional objectives. (Answers the question, "How good?")

Evaluation is much more comprehensive and inclusive term than measurement and testing is just one type of measurement. The term measurement is limited to quantitative descriptions of students, that is, results of measurement are always expressed in numbers. It does not include qualitative descriptions nor does it imply judgments concerning the worth or value of the obtained results. Evaluation on the other hand, may include both quantitative and qualitative descriptions of students. In addition, evaluation always includes value judgments concerning the desirability of the results.

The process of evaluation will be more effective when following principles are taken into consideration:

1. *Clearly specifying what is to be evaluated has priority in the evaluation process*: The effectiveness of evaluation depends as much as on a careful description of what to evaluate as it does in the technical qualities of the evaluation instruments used. Thus, specification of characteristics to be measured should precede the selection or development of evaluation devices when evaluating students' learning, this means clearly specifying the intended learning outcomes before selecting the achievement measures to use.
2. *An evaluation technique should be selected in terms of its relevance to the characteristics or performance to be measured*: Evaluation techniques are frequently selected on the basis of their objectivity, accuracy or convenience. Although these criteria are important, they are secondary to main criteria – Is this technique the most effective for measuring the learning or development to be evaluated? Each technique is appropriate for others. In testing students' achievements, for example, we need a close match between

the intended learning outcomes and the types of test items used.

3. *Comprehensive evaluation requires a variety of evaluation techniques:* No single type of instrument or procedure can assess the vast array of learning and development outcomes emphasized in an educational programme. Objective tests of achievement are useful for measuring knowledge, understanding and application outcomes but essay type tests and other written projects are needed to measure the ability to organize and express ideas. Observational techniques are needed to assess performance skills, various aspects of students' behaviour and self report techniques are useful for evaluating interests and attitudes. A complete picture of students achievements and developments requires continuing the results of many different techniques.
4. *Proper use of evaluation techniques requires an awareness of their limitations:* Evaluation techniques range from highly developed measuring instruments (Eg: Standardized aptitude test and achievement tests) to rather crude assessment devices (Eg: observational and self reporting techniques). Even our best educational and psychological measuring instruments, however, yield results that are subject to various types of measurement errors.
5. *Evaluation is a means to an end, not an end in itself:* The use of evaluation techniques implies that some useful purpose will be served and the user is clearly aware of this purpose. To blindly gather data about students and then file the information away in the hope that it will some day prove useful is a waste of time and effort. Evaluation is best viewed as a process of obtaining information on which to have decision.

Types of Testing and Evaluation Procedures

The role of the distinctive features of the evaluation process is the use of wide variety of procedures. These may be classified and described in many different ways depending on the frame of reference used. This classification is based on the basis of most useful application in the area of teaching.

Maximum Performance versus Typical Performance

In the first category are those procedures used to determine a person's abilities. Procedure for this type is concerned with How well an individual performs when he/she is motivated to obtain as high a score as possible. In short, the evaluation results indicate what an individual can do when he puts forth his best effort. Aptitude and achievement tests are included in this category. A distinction between these two types of tests is commonly mode in terms of the use of the results rather than the qualities of test themselves. An aptitude test is primarily designed to predict success in some future learning activity, while an achievement test is designed to indicate degree of success in some past learning activity.

The second category includes those designs to reflect a person's typical behaviour. Procedures of this type are concerned with what individuals will do rather than what they can do. Methods designated to evaluate interests, attitudes, adjustment and various personality traits are included in this category. Here, the emphasis is on obtaining representative responses rather than high scores. This type of evaluation involves various methods such as interviews, questionnaires, ratings and various other self reports and observational techniques. None of these techniques when used alone provide an adequate appraisal of typical behaviour but combined results of number of them enable the instructor to make fairly accurate judgments concerning students' progress.

Placement, Formative, Diagnostic and Summative Evaluation

Tests and other evaluation procedures can also be classified in terms of their functional role in classroom instruction.

Placement Evaluation

To determine student performance at the beginning of instruction *i.e.* measures entry behaviour

Formative Evaluation

to monitor learning progress during instruction, *i.e.* mid–term exam.

Diagnostic Evaluation

To diagnose learning difficulties during instruction, *i.e.* identifies causes of learning problems.

Summative Evaluation

to evaluate achievement at the end of instruction, *i.e.* measure end of course achievement.

Needless to say conventional evaluation fosters competition rather than collaboration. Formal assessment tends to be governed by 'what is easy to assess' rather than 'what is to be assessed'. Therefore, traditional written examination which has more transparency is used. Some of the types of tests which are frequently used are as follows:

Informal versus Standardized Tests

Informal tests are those constructed by classroom teacher, whereas those designed by specialists and administered, scored and interpreted under standard conditions are called standard tests.

Individual versus Group Tests

Some tests are administered on one–to–one basis whereas others can be administered to a group of individuals.

Mastery versus Survey Test

Some tests measure the degree of mastery of a limited set of specific outcomes whereas others measure a student's general level of achievement over a broad range of outcomes.

Supply versus Selection Tests

Some tests require examinee to supply the answer (*e.g.* essay test), whereas others require them to select the correct response from a given set of alternatives. (*e.g.* Multiple–choice test)

Speed versus Power Tests

A speed test is designed to measure the number of items an individual can complete in a given time, whereas a power test is designed to measure level of performance under ample power conditions. Power tests usually have the items arranged in order of increasing difficulty.

Objective versus Subjective Tests

An objective test is one in which equally competent scores will obtain the same score (*e.g.* Multiple choice test) whereas a subjective test is one where the scores are influenced by the opinion or judgment of the person doing the scoring (*e.g.* Essay test)

An adequate evaluation programme must include evaluation, development of desirable physical and mental health, social relationships, critical thinking appreciations, creative expression, interests and attitudes. Successful teaching calls for the competent use of evaluation of all three types of learning–cognitive, affective and psychomotor.

Steps in the Process of Evaluation

1. The major objectives of a curriculum should be developed cooperatively by the administrators and the teachers according to the needs of the students and the society they live in; some examples of major objectives are:
   (*a*) To develop an understanding of the diversification of agriculture.
   (*b*) To develop abilities basic to successful agriculture development.
   (*c*) To develop an understanding of various vocations those are possible in agriculture.
   (*d*) To develop a positive attitude towards farm family as a basic unit of agricultural development.
2. *Definition and classification of these objectives in terms of specific knowledge, abilities, skills and attitudes.*

   Major Objective
   - To develop abilities basic to successful agriculture development:

     Knowledge–to develop understanding regarding the role of agriculture for increased production.
   - To know the sources of various nutrients

     Abilities–to develop an ability to plan sustainable agriculture project.
   - to be able to plan low input agriculture cultivation.
   - To be able to produce low input sustainable varieties.

Skills–to develop skill regarding mushroom cultivation.

– To develop skill regarding production of value added products of bamboo.

Attitude–to develop a positive attitude towards low cost sustainable agriculture.

3. Selection of available tests or measures for each major objective:

   Example -

   Knowledge–objective test

   – Planning a lay- out for a kitchen garden.

   Ability–practical of budding, grafting, flower arrangement etc.

4. Construction of needed tests, scales and other techniques,

   Example:

   Preparation of–"True and false items"
   "Multiple choice items"
   "Short answer test items"

It is true that the main objectives of tests and examinations are mastery, diagnosis and reporting progress. In fact, the purpose of test should affect the kind of test given. Thus, a formal selection examination will normally cover a large number of skills and topics lightly and most questions should cover a large number of skills and topics lightly and most questions should be of middle difficulty level mastery tests on a particular unit which will be more intensive with several questions on each of a few topics lightly and most questions should be of middle difficulty level mastery tests on a particular unit which will be more intensive with several questions on each of a few topics to see whether they are mastered or not. Some tests might be relatively hard for all pupils, (*e.g.* Diagnostic tests). Some on the other hand would be lengthy and formal with elaborate marking schemes (*e.g.*: End of year examination).

Conclusion

Tests and evaluation play an important role in the instructional programme of an institution. Basically, they provide information that can be used in variety of educational decisions. The main emphasis in classroom evaluation however, is in decisions concerning students' learning and development.

Chapter 30

Training of Teachers for Qualitative Improvement in Teaching

Globalization has contributed towards rise in living standards, improvement in health, education and technology especially in the fields of communication and computers. The proposed WTO initiatives are expected to bring pressure on the universities worldwide into sharp focus. It is believed that Indian agricultural education would be significantly altered like higher education worldwide due to influecnes of WTO. The challenges facing Indian agricultural education are different than before. With the liberalization of economy, a number of multi-national companies with high tech agriculture are entering into agriculture and the competitiveness is increasing. The question before us is how to improve agricultural education to make graduates highly competitive nationally and internationally. There has been widespread consensus that quality of graduates in various disciplines of agricultural sciences does not match with the changing needs. With emerging job market scenario in front, introduction of subjects and methods of teaching and learning aligned fully with employability should form the core of the strategy for agricultural education. Furthermore, agricultural education curricula need to be redirected to address to national and international problems. Education has acquired a new dimension with phenomenal breakthrough in information and communication technologies. The emergence of virtual classroom, open learning systems and distance education has given new perspective to teaching and learning process. The challenges of education call for technological interventions to reap the fruits of emerging developments in education technology and instruction system design for which there is little awareness and no preparation in agricultural higher education Thus, attempts are urgently needed to create a system of teachers' education for equipping them to face the challenges of the hour. The paper aims to highlight the aspects of training for qualitative improvement in teaching.

Need for Training

Teachers training programmes in India had a glorious past but it has miserably failed (at least in recent years) to provide teachers, who can inculcate love for learning or co-curricular activities in students. Assessment of quality teaching and teachers should be a regular feature for improvement and attaining standards and excellence in agricultural education. Earlier for higher education no teacher training or education were considered necessary or desirable. This was due to the importance of the subject matter as compared to methodology in higher education. But it is now felt that teaching skill is also a significant contributor to the teacher's performance. Today most of the teachers are characterized by

- ☆ Poor surface knowledge of the subject, has specialized in and an extremely limited ability to apply the knowledge to a new situation.
- ☆ Poor language and poor communication skills
- ☆ Inability to meaningfully interact with the students to provide leadership.
- ☆ Lack of aptitude and consequently interest in the profession.
- ☆ Absence of desire to learn.

Teachers' talents and capabilities are directly correlated with the performance of students in term of their knowledge and skills development. In-service teacher education is increasingly seen as essential to the whole process of becoming an expert teacher and as a prerequisite for differentiation of tasks among teachers.

Studies on improving the quality of the training programmes have indicated that training activities in Africa covered aspects such as induction, workshops, self study etc. Most of the studies gave stress on the use of information and communication technology, action research, use of audio-visual aids (Berendt, 1989, El-Hares, 1994. Prakash& Kumar (2000) reported that teachers perceived highest training need in educational concepts and course design because most of the teachers were not exposed to these techniques through training, seminars and other interactions. Ahire (2005) found that top three areas in order of teachers' training need were curriculum design, teaching skills and teaching methods. Jethi (2006) in her study at G.B. Pant University of Agriculture and Technology to found that students rated teachers low on use teaching methods and content organization.

Strategy for Training

Assessment of quality Teaching and teachers should be a regular feature for improvement for attaining standards and excellence in Agricultural Education. Teachers' talents and capabilities are directly correlated with the performance of students in term of their knowledge and skills development. A well-defined and transparent career management and advancement policy need to be evolved for faculty in order to motivate them to enhance the job satisfaction and eliminating stagnation. Teacher education cannot provide for expert teachers, but it must be possible to contribute to the development of personalities who are able and willing to become experts. It is difficult to say what an expert teacher should know and be able to do, because of changes in education and society. Therefore, it is perhaps better to pay more attention to the conditions for the process of becoming an expert teacher than to the determining of the requirements a teacher should meet. During this process special attention needs to be paid to the relationship between students and teachers. In-service teacher education is increasingly seen as essential to the whole process of becoming an expert teacher and as a prerequisite for differentiation of tasks among teachers. Less than in the past there is a clear distinction between pre- service and in-service teacher education by having described elements of competent teaching, the everyday teaching practice and the professionalization of teachers.

Training of teachers can best be done through micro-teaching. Microteaching refers to the technique of training teachers in teaching skills through microteaching sessions lasting not more than five to seven minutes with peers as learners. Such sessions help to integrate skills of teaching one at a time. Trainer identifies a skill and demonstrates before the teacher trainees. Teacher trainees plan and enact the lesson. Recorded teaching is played for analysis and feedback for the teachers to re-plan, re teach, replay, reanalyse and feedback. This helps the new as well as experienced teachers to improve their teaching in a comfortable situation. Recording in audio or video form is essential for reviewing by the teacher. Feedback is provided in the basis of observation on pre-structured Performa. This is an important tool for research on classroom teaching behaviour and interaction. Curriculum design is foundation for teaching. There is need to continuously revise curriculum according to the needs of the market. Participatory curriculum design should be practiced.

Preparation and Exchange of Standardized Teaching

Normally the contents and methodology developed by individual teachers remain with them and seldom shared. If the teachers retire or shift jobs their unique experiences go away with them. The experience and wisdom of teachers must be preserved in planned manner to incorporate changes and refine courses. Thus every senior teacher may be asked to record their lecture notes along with aids and approaches used by them on computer so as to standardize the teaching objectives, content and methodology. Such packages clearly indicate the course outline, specific behavioural objectives, lecture notes along with photocopies of related reference materials and list of assignments along with criteria of evaluation. The whole strategy can be discussed with colleagues and even shared with experts outside for possible changes. Such teaching packages can be exchanged with teaching faculties in the other universities within the country.

Experiential Learning

Now that professionalism calls for performance and problem solving students must be prepared to perform. This is possible only when students work in simulated environment with similar tools and techniques, which are to be used in real life. This calls for more of hands on experiences and discussion in classes, regular interaction with industries and employers to get ready for the desired roles. Teaching is not about delivering lecture but making learners aware of the concepts and practices in a manner that they get interested to explore the literature and learn on their own. It is a participatory approach in which students can directly apply what they are learning. In order to achieve this, professors must first design the active learning exercises and then facilitate the learning. A teacher should now make the transition from a teacher-centered approach, where they are seen as a transmitter of knowledge, to a more student-centered approach where they become a facilitator of student learning. Research indicates that active learning increases motivation and attendance. It reduces competition and isolation in the classroom by involving all students (including marginalized students) and having everyone learn to work together cooperatively.Thus, all the materials to be learnt should be made available to the students in soft or hard copies so that students can come prepared for discussion, analysis and problem solving.

Distance Learning

Teachers must get ready to write teaching packages for distance learners. This will help them learn the ways to prepare on line materials and in turn such materials can be used by face-to-face learners. Besides, students can be encouraged to use on-line materials available on internet.

Summing Up

- ☆ There is an ongoing revolution in the field of teaching and learning especially in the professional fields to the extent that classroom teaching is becoming obsolete in many professional universities of the world. Preparing graduates to diagnose, solve problems and adjust with different types of work conditions calls for use of innovative methods to provide opportunity to apply and test skills.
- ☆ The experience and wisdom of teachers must be preserved in planned manner to incorporate changes and refine courses. Thus every senior teacher may be asked to record their lecture notes along with aids and approaches used by them on computer Such teaching packages can be discussed with colleagues and even shared with experts outside for possible changes.

- ☆ Teachers are required to improve upon their teaching skills by incorporating use of information technologies for developing lesson plans, gathering current relevant information, keeping students' records and developing objective interactive tests and simulation games. Teachers should have knowledge of different software available, apart from computers what other equipments and peripherals can be used to prepare multi-media modules for teaching. In this information age teachers should have the knowledge of database available in agriculture and allied fields like AGRICOLA and CAB abstracts are available online. Teachers may enable students by providing online address pertinent to specific fields. Teachers should not only know concepts, course design, media or teaching process but also how to evaluate. There are an array of approaches, methods and ways of constructing tests.
- ☆ Each SAU needs centre for In-house training of teachers in teaching skills. Beginning with new teachers, the emphasis would be to impart skills of effective teaching and dissemination of best teaching practices to all the teachers. Besides, support would be provided to teachers in preparing teaching packages in digital form, guidance in preparation of portfolio of interested teachers, formation of quality circle and conducting action research, facility for video recording of teaching to interested teachers. In order to consolidate the gains on continuous basis and learn from the theories of education, each SAU needs to establish an educational technology cell to improve teaching through training and dissemination of best teaching practices.

Bibliography

Ahire, M.C. (2005).Classroom Teaching Performance and Training Preferences of University Teachers. Unpublished Ph.D.thesis, Dept.-Agriculture Communication, G.B.Pant University of Agriculture and Technology.

Ahire C Milind (2005) Classroom Teaching Performance and Training Preferences of University Teachers. Unpublished Ph.D Thesis, G. B. Pant University of Agriculture and Technology.

Alden, D (1999), Experience with Scripted Role – Play in Environmental Economics, Journal of Economic Education, Spring 1999.

Berendt B (1989) Improving Teaching and Learning in African Universities: Staff (Faculty) Development Programmes as a Field of Cooperation between European and African Universities, Higher Education in Europe, 14 (2), pp.69-75.

Bonwell, C.C., and J.A. Eison. 1991. *Active learning: Creating excitement in the classroom.* ASHE-ERIC Higher Education Report No. 1. Washington, DC: George Washington University.

Brent, R., and R.M. Felder. 1992. Writing Assignments – Pathways to Connections, Clarity, Creativity. *College Teaching.* 40(2):43 - 47.

Campbell, W. E., and K.A. Smith (Eds.). 1997. *New paradigms for college teaching*. Edina, MN: Interaction Book Company.

Craig, R.L. and L.R.Bittel (Eds) (1967). Training and Development Hand Book, New York, McGraw Hill, Sponsored by American Society of Training and Development

El-Hares, H (1994) University Teaching Network: The Arab Network for staff Development. In UNESCO, Higher Education Staff Development: Directions for the twenty first century, UNESCO, Paris, pp. 93-106.

FAO (1990), Making Learning Easier: A guide for Improving Education/Training Materials. Rome, FAO

Felder, R.M. and R. Brent. 1999. How to Improve Teaching Quality. *Quality Management Journal*, 6(2): 9-21.

Freire, P. (1972), Pedagogy of the Oppressed, Harmonds Mouth, Penguin.

Freire, P. (1974), Education: The Practice of Freedom. London, Writes and Readers Publishing Cooperative.

Freeman, M. A. and Capper, J. M, an Anonymous Asynchronous Web – based Role – Play. (1998)

Gronlund, N.E.1991. *How to write and use instructional objectives* (4th ed). New York: Macmillan.

Hansra, B.S. and Kumar, B. (1996). Training Methodology for Human resource Development, New Delhi, Classical Publishing

http://www.bus.uts.edu.au/fin&econ/staff/markf/roleplay.html

http://ausweb.scu.edu.au/aw01/papers/refereeed/ip/paper.html

Iffner, P. L. and A.F. Douds (1988) Train the Teacher-Practical Skills that work, HRD Press, Massachusetts, USA

Ip, A, Linser, R and Naidu, S. Simulated worlds: Rapid Generation of Web – based Role – Play (2001)

I.L.O. (1986). Vocational Training, Glossary of selected Terms, International Labour Office, Geneva

Johnson, D.W., R.T. Johnson, and K.A. Smith. 1998. *Active learning: Cooperation in the college classroom*, 2d ed. Edina, MN: Interaction Press.

Jules N. Pretty (1995) Participatory Learning and Action: A Teacher's Guide.II ED Participatory Methodology Series.

Jethi Renu (2006) Perception of Teachers and Students towards Teachers' Evaluation by Students: A Study at G. B. Pant University of Agriculture and Technology, Unpublished Ph.D Thesis, Pantnagar

Kumar,B. and Hansra,B.S. (2000). Extension Education for Human Resource development, New Delhi, Concept Publishing Co.

Kumar,B. and Kashyap,S.K.(2005). Management of Training : A Guide for Trainers, NATP(TOE in Agricultural Information and communication), GB Pant University of Agriculture and Technology, Pantnagar

Mehta, P (1983). New Extension and Training Strategies, Indian Journal of Extension Education, 19(1&2), pp.-1-14

McKeachie, W. 1999. *Teaching tips*, 10th ed. Boston: Houghton Mifflin.

Meyers, C., and T.B. Jones. 1993. *Promoting active learning*. San Francisco: Jossey-Bass.

Millis, B.J., and P.G. Cottell, Jr. 1998. *Cooperative learning for higher engineering faculty*. Phoenix: Oryx Press.

NISE (National Institute for Science Education). 1997. *Collaborative learning: Small group learning page. <http://www.wcer.wisc.edu/nise/cl1/>*

Nadler, L (1994) The Handbook of Human Resource Development, John Wiley and sons

Paralikar, K. R. (1990), "The Art of Teaching Home Science", Ekvira Publications, Baroda. 110p.

Prakash and Kumar (1996). Training of Agricultural University Teachers in Education Technology, University News 34(5).

Prakash D and Kumar B (2000) Training Needs of Agricultural University Teachers in Educational Technology. Teacher Education in India, Selections from University News-5 (122-127)

Prakash, D. (1995). Training needs of Educational Technology, A Study at G.B. Pant University of Agriculture and Technology.Unpublished M.Sc..thesis, Dept.-Agriculture

Communication, G.B.Pant University of Agriculture and Technology.

Prasad, C. (1981). Elements of structure and terminology of Agricultural Education in India, Paris, France, UNESCO (1981) 82pp In: World Agricultural Economics and Rural sociology,1981,Vol 24(7) pp.530.

Roob, R.T, Swanson, B.E Wentling T.L., and clark, D.D., (1991), Improving Training Quality: A Teacher's Guide to Evaluation. Rome, FAO

Rosemary S Caffarella (1987) Program Development and Evaluation Resource Book for Teachers

Singh, P. N. (1989). Changing Roles of Teachers, Indian Journal of training and Development, July-September,1984

Singh, P. N. (1989). Training for Management Development, Forum for Asian Managers and Indian Society for Training and Development

Singh, P.N. and Miranda, S. (1986). Role of a Teacher, Indian Journal of training and Deevlopment, January-March, 1986,pp 30-40

Strayton, R. (1986) Organization of courses and Conferences in ILO, Chapter 21, International Labour Office, Geneva

Sundramurthy, S. (1986) Teachers' role effectiveness, Indian Journal of Training and Development, September-October,1986

Wadhera, K (1979) Training Manual in Helping Profession, Vishwa Yuwak Kendra, New Delhi

Bates, A.W. 1995. *Technology, Open Learning and Distance Education.* London: Routledge.

Dede, C.1996. "Emerging Technologies in Distance Education for Business." *JOURNAL OF EDUCATION FOR BUSINESS 71*, no. 4 (March-April 1996): 197-204.

Gerard G. 1997. Selecting Instructional Media for Distance Education, http://www.icdi.wvu.edu/rta/symp97/gerard.htm, Department of Special Education and Rehabilitation, Utah State University Logan, Utah.

Hodgson, B. 1993. *Key Terms and Issues in Open Learning.* London: Kogan Page.

Kelsey, T. W. and Mincemoyer, C. C., 2001. Exploring the potential of inservice training through distance education. Journal of Extension {on-line]. 39 (2). Available: http://www.joe.org/joe/2001april/rb7.html

Kumar and Kashyap 2004. Management of Training: A Guide for Trainers; NATP (TOE) in Agricultural Information and Communication; G.B. Pant University of Agriculture and Technology, Pantnagar, Uttarakhand.

Mills, R. 2000. The Role of Study Centres in Open and Distance Education: Glimpse of the Future."In Supporting Students in Open and Distance Learning. London:Kogan Page.

Mishra A.K. and Bartram J.2002.Perspectives on Distance Education – Skills Development through Distance Education. The Commonwealth of Learning.

Mishra, P.2008. Designing A Training Module for Extension Trainers: A Study In Uttarakhand. Unpublished Ph.D., Thesis, Department of Agriculture Communication, G.B. Pant University of Agriculture and Technology, Pantnagar.

Mishra, D.C. 1990. New Directions in Extension Training; Directorate of Extension, Department of Agriculture and Co-operation; Ministry of Agriculture, New Delhi.

Moore and Thompson, 1990.Verduin and Clark, 1991 Moore. M.G. and Thompson, M.M., with Quigley, A.B., Clark, G.C., and Goff, G.G. 1990. The effects of distance learning: A summary of the literature. Research Monograph No. 2. University Park, PA: The Pennsylvania State University, American Center for the Study of Distance Education. (ED 3 30 321).

Neuhauser, C. 2002. Learning style and effectiveness of online and face-to-face Instruction. *The American Journal of Distance Education, 16,* 99-113.

Panda, S. and S. Chaudhary, 2001."Telelearning and Telelearning Centres in India." In *Telecentres: Case Studies and Key Issues.* Vancouver: The Commonwealth of Learning.

Pandey, R.K. 1992. Need assessment of KVK trainers in UP in training management. Unpublished M.Sc.(Ag.), Thesis, Department of Agriculture Communication, G.B. Pant University of Agriculture and Technology, Pantnagar.

Rajput,A. and Kumar,B.2011. Designing Strategy for KVK Trainers' Training in Trainers' Skills through Distance Learning – A Study in North India Unpublished Ph.D., Thesis, Department of Agriculture Communication, G.B. Pant University of Agriculture and Technology, Pantnagar.

Simpson, O.2000. Supporting Students in Open and Distance Learning. London: Kogan Page.

Smith, P.L. and T.J. Ragan.1999. *Instructional Design*. New York: Macmillan Publishing Company.

Sreekumar, S.S. 1998. Study of Skills Development through Distance Education. *Indian Journal of Open Learning*, 1998, 7(2), 199-204.

UNESCO 1996b. *Learning: the Treasure Within*. Report to UNESCO of the International Commission on Education or the Twenty-first Century, Paris: UNESCO.

Yadav, B. and Verma, T. 1998. Assessing trainers' view about various aspects of training. *Maharashtra Journal of Extension Education*.17:248-255.

Govind Ballabh Pant University of Agriculture and Technology, Pantnagar: Information and Communication Technology Policy, 2006

K.B.Powar: Information Technology and Higher Education: The Indian Context, Information Technology in Higher Education, Association of Indian Universities, 2000

S. Sudhakar Babu: Technological Advances and Role of ICT in Teacher Training and Higher Education, University News, Association of Indian Universities, June25-July01,2007

Report of the task force of the International Council on Distance Education (ICDE): Implications for ICDE and the Distance Learning Community, 1996

Lindley, W.I. (1998) The Relevance of Higher Education in Agriculture and Rural Development. Speech delivered at World Conference on Higher Education; Higher Education in Twenty first Century; Vision and Action. UNESCO, Paris, 5 9 October [Online] Available: http://www.unesdoc.unesco.org/images/0011/001170/007075e. pdf Pretty, J.N. (1995).

Regenerating Agriculture: Policies and Practice for Sustainable and Self-Reliance, London, England: Earthscan Publications, 320.

Reeves, T. (1998) Sustainable Intetensification of Agriculture. CIMMYT. Apdo. Postal 6-641, 06600 Mexico. [online] Available: http"//www.cimmyt.cgiar.org/whatiscimmy/Sustint.htm

Rodriguez, Lylian and John Kornerup Bang (1999) Capacity Building for Sustainable Rural Development: Education, a Priority. Proceedings of a Workshop March 22–26, 1999 Tune Landboskole Denmark Organizer: The Danish Agricultural and Rural Development Advisers' Forum [Online] Available http://www.husdyr.kvl.dk/php/tune99/27-Rodrigues.htm

Rogers, A (1996). Participatory Training: Using Critical Reflection On Experience in Agricultural Extension Training. FAO Training for Agriculture and Rural Development, Economic and Social Development Series, No. 54.Smith, Nicholas (undated) Seminar Review: No. 7 Sustainable Agriculture–Myth of Reality? [online] Available : http://www.science.plym.ac.uk/departments/geography/ggy384/susagr2.htm

The World Bank (1999) World Development Report 1998/99: Knowledge for Development-Summary. The International Bank for Reconstruction and Development. 1818H Street, N.W. Washignton D.C. U.S.A.